Learning Jupyter 5
Second Edition

Explore interactive computing using Python, Java, JavaScript, R, Julia, and JupyterLab

Dan Toomey

BIRMINGHAM - MUMBAI

Learning Jupyter 5
Second Edition

Commissioning Editor: Pravin Dhandre
Acquisition Editor: Tushar Gupta
Content Development Editor: Chris D'cruz
Technical Editor: Suwarna Patil
Copy Editor: Safis Editing
Project Coordinator: Nidhi Joshi
Proofreader: Safis Editing
Indexer: Priyanka Dhadke
Graphics: Tom Scaria
Production Coordinator: Arvindkumar Gupta

First published: November 2016
Second edition: August 2018

Production reference: 1290818

Published by Packt Publishing Ltd.
Livery Place
35 Livery Street
Birmingham
B3 2PB, UK.

ISBN 978-1-78913-740-8

www.packtpub.com

`mapt.io`

Mapt is an online digital library that gives you full access to over 5,000 books and videos, as well as industry leading tools to help you plan your personal development and advance your career. For more information, please visit our website.

Why subscribe?

- Spend less time learning and more time coding with practical eBooks and Videos from over 4,000 industry professionals

- Improve your learning with Skill Plans built especially for you

- Get a free eBook or video every month

- Mapt is fully searchable

- Copy and paste, print, and bookmark content

PacktPub.com

Did you know that Packt offers eBook versions of every book published, with PDF and ePub files available? You can upgrade to the eBook version at `www.PacktPub.com` and as a print book customer, you are entitled to a discount on the eBook copy. Get in touch with us at `service@packtpub.com` for more details.

At `www.PacktPub.com`, you can also read a collection of free technical articles, sign up for a range of free newsletters, and receive exclusive discounts and offers on Packt books and eBooks.

Contributors

About the author

Dan Toomey has been developing application software for over 20 years. He has worked in a variety of industries and companies, in roles from sole contributor to VP/CTO-level. For the last few years he has been contracting for companies in the eastern Massachusetts area. Dan has been contracting under Dan Toomey Software Corp. Dan has also written *R for Data Science*, *Jupyter for Data Sciences*, and the *Jupyter Cookbook*, all with Packt.

About the reviewer

Juan Tomás Oliva Ramos is an environmental engineer from University of Guanajuato, Mexico, with a master's degree in administrative engineering and quality. He has more than 5 years of experience in management and development of patents, technological innovation projects, and technological solutions through the statistical control of processes. He has been a teacher of statistics, entrepreneurship, and technological development since 2011. He has developed prototypes via programming and automation technologies for the improvement of operations, which have been registered for patents.

I want to thank God for giving me wisdom and humility to review this book. I thank Packt for giving me the opportunity to review this amazing book and to collaborate with a group of committed people. I want to thank my beautiful wife, Brenda, our two magic princesses (Maria Regina and Maria Renata), and Angel Tadeo, all of you, give me the strength, happiness, and joy to start a new day. Thanks for being my family.

Packt is searching for authors like you

If you're interested in becoming an author for Packt, please visit authors.packtpub.com and apply today. We have worked with thousands of developers and tech professionals, just like you, to help them share their insight with the global tech community. You can make a general application, apply for a specific hot topic that we are recruiting an author for, or submit your own idea.

Table of Contents

Preface

Learning Jupyter discusses using Jupyter to record your scripts and produce results for data analysis projects. Jupyter allows data scientists to record their complete analysis process, much in the same way that other scientists use a lab notebook for recording tests, progress, results, and conclusions. Jupyter works in a variety of operating systems, and this book covers the use of Jupyter in Windows and macOS, along with the various steps necessary to enable your specific needs. Jupyter supports a variety of scripting languages by the addition of language engines, so the user can use their particular script in a native fashion.

Who this book is for

This book is written for readers who wants to portray software solutions to others in a natural programming context. Jupyter provides a mechanism to execute a number of different languages and stores the results directly for display, as if the user ran those scripts on their own machine.

What this book covers

Chapter 1, *Introduction to Jupyter*, investigates the various user interface elements available in a notebook. We will learn how to install the software on a macOS or a PC. We will expose the notebook structure. We will see the typical workflow used when developing a notebook. We will walk through the user interface operations available in a Notebook. And lastly, we will see some of the configuration options available to advanced users for their notebook.

Chapter 2, *Jupyter Python Scripting*, walks through a simple notebook and the underlying structure. Then, we will see an example of using pandas and looked at a graphics example. Finally, we will look at an example using random numbers in a Python script.

Chapter 3, *Jupyter R Scripting*, adds the ability to use R scripts in our Jupyter Notebook. We will add an R library that's not included in the standard R installation, and we will make a Hello World script in R. We will then see R data access built-in libraries and some of the simpler graphics and statistics that are automatically generated. We will use an R script to generate 3D graphics in a couple of different ways. We will then perform a standard cluster analysis (which I think is one of the basic uses of R) and use one of the forecasting tools. We will also build a prediction model and test its accuracy.

Chapter 4, *Jupyter Julia Scripting*, adds the ability to use Julia scripts in our Jupyter Notebook. We will add a Julia library that's not included in the standard Julia installation. We will see the basic features of Julia in use, and also outline some of the limitations that are encountered using Julia in Jupyter. We will display graphics using some of the available graphics packages. Finally, we will see parallel processing in action, a small control flow example, and how to add unit testing to your Julia script.

Chapter 5, *Jupyter Java Coding*, explains how to install the Java engine into Jupyter. We will see examples of the different output presentations available from Java in Jupyter. Then, we will investigate using optional fields. We will see what a compile error looks like in Java in Jupyter. Next, we will see several examples of lambdas. We will use collections for several purposes. Lastly, we will generate summary statistics for one of the standard datasets.

Chapter 6, *Jupyter JavaScript Coding*, shows how to add JavaScript to our Jupyter Notebook. We will see some of the limitations of using JavaScript in Jupyter. We will look at examples of several packages that are typical of Node.js coding, including graphics, statistics, built-in JSON handling, and creating graphics files with a third-party tool. We will also see how multithreaded applications can be developed using Node.js under Jupyter. Lastly, we will use machine learning to develop a decision tree.

Chapter 7, *Jupyter Scala*, explains how to install Scala for Jupyter. We will use Scala coding to access large datasets. We will see how Scala can manipulate arrays. We will generate random numbers in Scala. There are examples of higher-order functions and pattern matching. We will use case classes. We will see examples of immutability in Scala. We will build collections using Scala packages, and we will look at Scala traits.

Chapter 8, *Jupyter and Big Data*, discusses using Spark functionality via Python coding for Jupyter. First, we will install the Spark additions to Jupyter on a Windows machine and a macOS machine. We will write an initial script that just read lines from a text file. We will go further and determine the word count in that file. We will add sorting to the results. There is a script to estimate pi. We will evaluate web log files for anomalies. We will determine a set of prime numbers, and we will evaluate a text stream for some characteristics.

Chapter 9, *Interactive Widgets*, explains how to add widgets to our Jupyter installation. We will use the interact and interactive widgets to produce a variety of user input controls. We will then look at the widgets package in depth to investigate some of the available user controls, properties available in the containers, and events that can be emitted from the controls, and we'll how to build containers of controls.

Chapter 10, *Sharing and Converting Jupyter Notebooks*, covers how to share notebooks on a Notebook server. We will add a notebook to a web server distribute it using GitHub. We will also look into converting our notebooks into different formats, such as HTML and PDF.

Chapter 11, *Multiuser Jupyter Notebooks*, shows how to expose a notebook so that multiple users can use a Notebook at the same time. We will see an example of the sharing error occurring. We will install a Jupyter server that addresses the problem, and we will use Docker to alleviate the issue as well.

Chapter 12, *What's Next?*, looks into some ideas that may be incorporated into Jupyter in the future.

To get the most out of this book

The steps in this book assume you have a modern Windows or macOS with internet access. There are several points where you will need to install software, so you will need administrative privileges on the machine to do so.

The expectation is you have one or more favorite implementation languages you wish to use on Jupyter.

Download the example code files

You can download the example code files for this book from your account at www.packtpub.com. If you purchased this book elsewhere, you can visit www.packtpub.com/support and register to have the files emailed directly to you.

You can download the code files by following these steps:

1. Log in or register at www.packtpub.com.
2. Select the **SUPPORT** tab.
3. Click on **Code Downloads & Errata**.
4. Enter the name of the book in the **Search** box and follow the onscreen instructions.

Once the file is downloaded, please make sure that you unzip or extract the folder using the latest version of:

- WinRAR/7-Zip for Windows
- Zipeg/iZip/UnRarX for Mac
- 7-Zip/PeaZip for Linux

The code bundle for the book is also hosted on GitHub at `https://github.com/PacktPublishing/Learning-Jupyter-5-Second-Edition`. In case there's an update to the code, it will be updated on the existing GitHub repository.

We also have other code bundles from our rich catalog of books and videos available at `https://github.com/PacktPublishing/`. Check them out!

Conventions used

There are a number of text conventions used throughout this book.

`CodeInText`: Indicates code words in text, database table names, folder names, filenames, file extensions, pathnames, dummy URLs, user input, and Twitter handles. Here is an example: "The default filename, `untitled1.txt`, is editable."

A block of code is set as follows:

```
var mycell = Jupyter.notebook.get_selected_cell();
var cell_config = mycell.config;
var code_patch = {
    CodeCell:{
      cm_config:{indentUnit:2}
    }
 }
cell_config.update(code_patch)
```

Any command-line input or output is written as follows:

```
jupyter trust /path/to/notebook.ipynb
```

Bold: Indicates a new term, an important word, or words that you see onscreen. For example, words in menus or dialog boxes appear in the text like this. Here is an example: "There are three tabs that are displayed: **Files**, **Running**, and **Clusters**."

Warnings or important notes appear like this.

 Tips and tricks appear like this.

Get in touch

Feedback from our readers is always welcome.

General feedback: Email `feedback@packtpub.com` and mention the book title in the subject of your message. If you have questions about any aspect of this book, please email us at `questions@packtpub.com`.

Errata: Although we have taken every care to ensure the accuracy of our content, mistakes do happen. If you have found a mistake in this book, we would be grateful if you would report this to us. Please visit `www.packtpub.com/submit-errata`, selecting your book, clicking on the Errata Submission Form link, and entering the details.

Piracy: If you come across any illegal copies of our works in any form on the Internet, we would be grateful if you would provide us with the location address or website name. Please contact us at `copyright@packtpub.com` with a link to the material.

If you are interested in becoming an author: If there is a topic that you have expertise in and you are interested in either writing or contributing to a book, please visit `authors.packtpub.com`.

Reviews

Please leave a review. Once you have read and used this book, why not leave a review on the site that you purchased it from? Potential readers can then see and use your unbiased opinion to make purchase decisions, we at Packt can understand what you think about our products, and our authors can see your feedback on their book. Thank you!

For more information about Packt, please visit `packtpub.com`.

Introduction to Jupyter

1

Jupyter is a tool that allows data scientists to record their complete analysis process, much in the same way other scientists use a **Lab Notebook** to record tests, progress, results, and conclusions.

The Jupyter product was originally developed as part of the IPython project. The IPython project was used to provide interactive online access to Python. Over time, it became useful to interact with other data analysis tools, such as R, in the same manner. With this split from Python, the tool grew into its current manifestation of Jupyter. IPython is still an active tool that's available for use. The name Jupyter itself is derived from the combination of Julia, Python, and R.

Jupyter is available as a web application from a number of places. It can also be used locally over a wide variety of installations. In this book, we will be exploring using Jupyter on a macOS and a Windows PC, as well as over the internet with other providers.

With Jupyter 5.0, there were significant enhancements for the following:

- Cell tagging
- Customizing keyboard shortcuts
- Copying and pasting cells between Notebooks
- A more attractive default style for tables

In this chapter, we will cover the following topics:

- First look at Jupyter
- Installing Jupyter
- Notebook structure
- Notebook workflow
- Basic Notebook operations
- Security in Jupyter
- Configuration options for Jupyter

First look at Jupyter

Here is a sample opening page when using Jupyter (this screenshot is on a Windows machine):

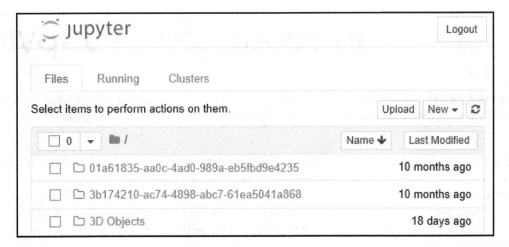

You should get yourself acquainted with the environment. The Jupyter user interface has a number of components:

- The product title, Jupyter, in the top left (as expected). The logo and the title name are clickable and will return you to the Jupyter Notebook home page.
- There are three tabs which are displayed: **Files**, **Running**, and **Clusters**:

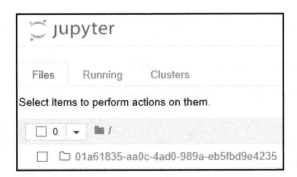

- The **Files** tab shows the list of files in the current directory of the page (described later on in this section).

- The **Running** tab presents another screen, which shows the currently running processes and Notebooks. The drop-down lists for **Terminals** and **Notebooks** are populated with their running members:

- The **Clusters** tab presents another screen which displays a list of available clusters. This topic is covered in a later chapter:

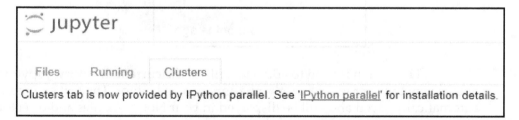

- In the top right corner of the screen, are three buttons: **Upload**, **New** (menu), and a Refresh notebook list button.
- The **Upload** button is used to add files to the Notebook space. You may also just drag and drop as you would when handling files. Similarly, you can drag and drop Notebooks into specific folders as well.

- The menu with **New** at the top presents a further menu of the Notebook for the different Notebook engines that have been installed (I had installed Jupyter earlier these are not default values) **Javascript (Node.js)**, **Julia 0.6.1**, **Python 2** (which will not be covered in this book), and **Python 3**. The additional **Other** menu items are **Text File**, **Folder**, and **Terminal**:

- The **Text File** option is used to add a text file to the current directory. Jupyter will open a new browser window for you, running a text editor. The text entered is automatically saved and will be displayed in your Notebook files and directory display:

The default filename, `untitled1.txt`, is editable. Note that the filename corresponds with the title given to the Notebook.

- The **Folder** option creates a new folder with the name **Untitled Folder**. Remember that all of the file and folder names are editable:

- The **Terminals** option is used to open a new Terminal (command) window. The resulting display on a Windows machine looks as follows:

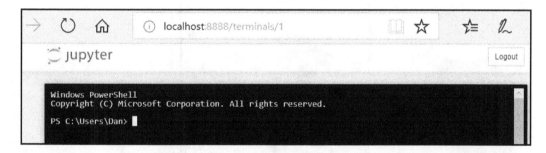

- The **Python 3** option is used to start a new Python 3 Notebook. The interface looks like it does in the following screenshot. You have full file editing capabilities for your script, including saving as a new file. You also have a complete working IDE for your Python script:

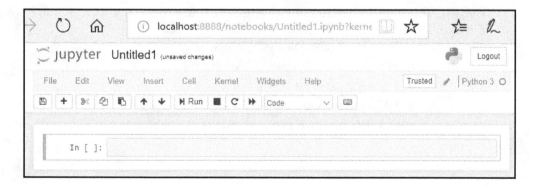

Note, like the **Text File** and **Folder** option, you have created a Python script file in your Notebook and it is running! (You can see this in the home page display of Jupyter):

- The Refresh notebook list button is used to update the display. It's not really necessary as the display is reactive to any changes in the underlying file structure.
- At the top of the **Files** tab is a checkbox, a drop-down menu, and a **Home** button.
- The checkbox is used to toggle all the checkboxes in the items list.
- The drop-down menu presents a list of the choices available, that is, **Folders**, **All Notebooks**, **Running**, and **Files**, as shown in the following screenshot:

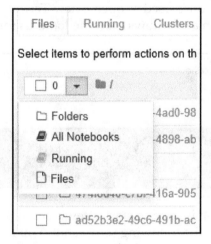

- The **Folders** selection will select all the folders in the display and present a count of the folders in the small box.
- The **All Notebooks** selection will change the count to the number of Notebooks and provide you with five options:
    - **Duplicate** (the selected Notebooks)
    - **Shutdown** (the selected Notebooks)
    - **View** (the selected Notebooks)
    - **Edit** (the selected Notebooks)
    - **Delete** (the trash can icon; the selected Notebooks)

- You can see them in the following screenshot:

- The **Running** selection will select any running scripts in the display and update the count to the number selected:

- The **Files** selection will select all of the files in the Notebook display and update the count accordingly.
- The **Home** button brings you back to the home screen of the Notebook.
- On the left-hand side of every item is a checkbox, an icon, and the item's name:

- The checkbox is used to build a set of files to operate upon.
- The icon is indicative of the type of item. In this case, all of the items are folders.
- The name of the item corresponds to the name of the object. In this case, the filenames are as they are when used on the disk.

Installing Jupyter

Jupyter requires Python to be installed (it is based on the Python language, after all). There are a couple of tools that will automate the installation of Jupyter (and optionally Python) from a GUI. In this case, we are showing you how to install Jupyter using Anaconda, which is a Python tool for distributing software.

First of all, you have to install Anaconda. It is available on Windows and macOS environments. Download the executable from https://www.continuum.io/ (the company that produces Anaconda) and run it to install Anaconda. Be sure to select the version of Anaconda using Python 3.x versus Python 2.x. The software provides a regular installation setup process, as shown in the following screenshot:

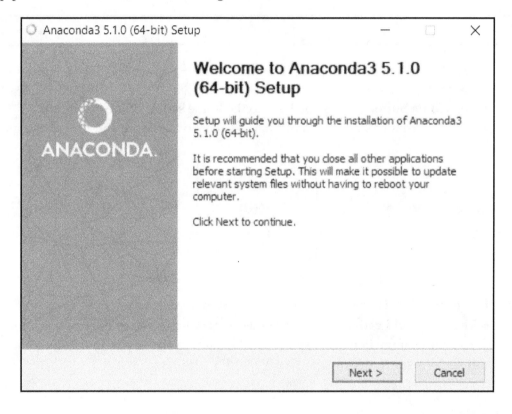

The installation process goes through the regular steps of making you agree to the distribution rights license:

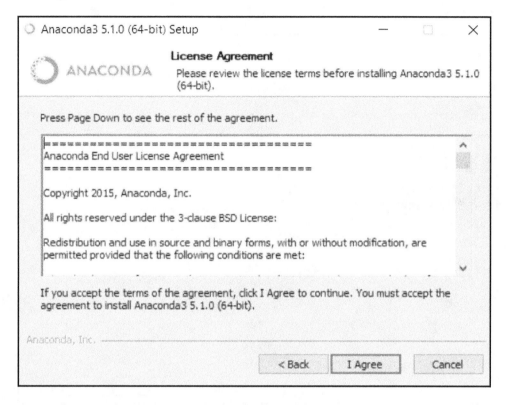

The standard Windows installation allows you to decide whether all users on the machine can run the new software or not. If you are sharing a machine with different levels of users, then you can decide upon the appropriate action:

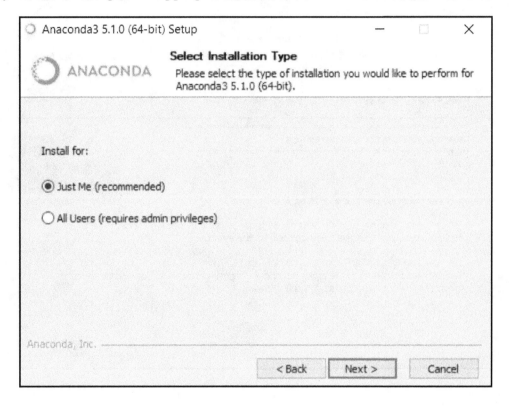

After clicking on **Next**, it will ask for a destination for the software to reside (I almost always keep the default paths):

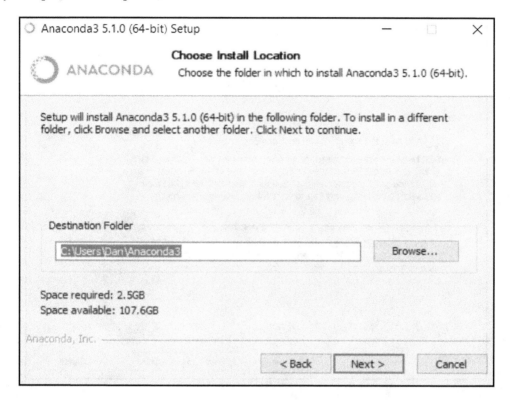

Anaconda will also adjust your file paths to make Anaconda accessible at all points on your machine via the next dialog box, as follows:

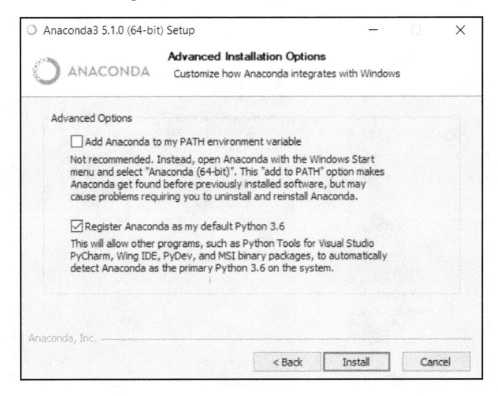

The installation will then begin. This may take a while, depending on your machine configuration and network access:

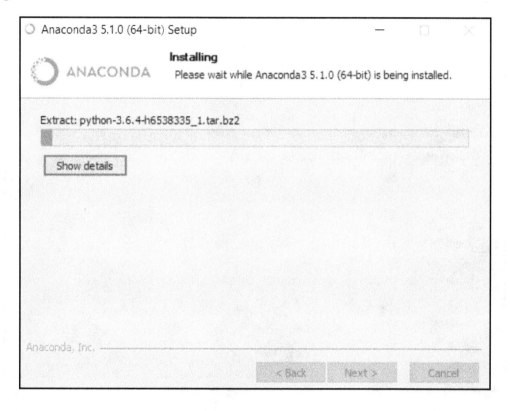

You will eventually get to the **Installation Complete** screen, as follows:

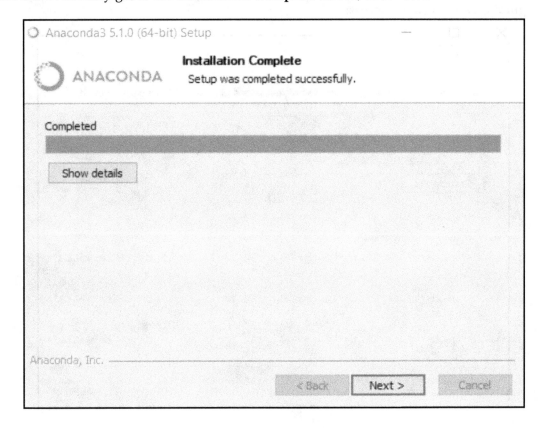

On Windows, Anaconda takes advantage of the semi built-in aspects of the Visual
Development Environment to access Windows services natively. It asks for permission to
do so with the following dialog:

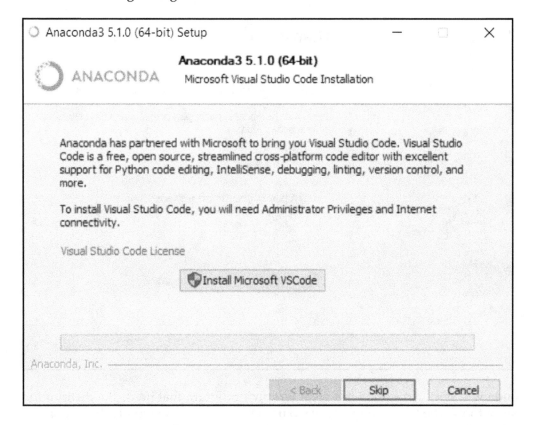

And now we have truly installed Jupyter:

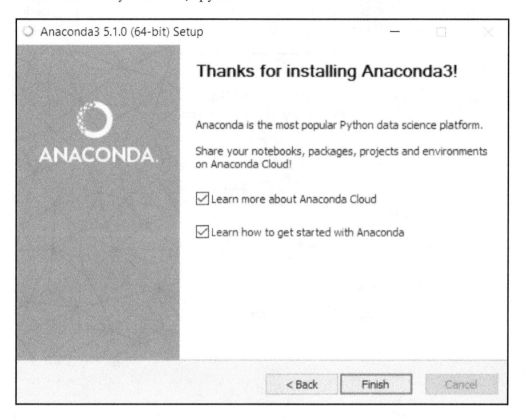

Anaconda will start. Anaconda is a great wrapper program that holds the distribution for a number of tools. The tool of importance to us is Jupyter. The Anaconda display shows the available tools, whether they need to be installed, and a starting place for each.

You can get to Jupyter directly by using the > `jupyter notebook` command from a Terminal window.

If we select Jupyter from the Anaconda screen, we will start Jupyter in a new browser window:

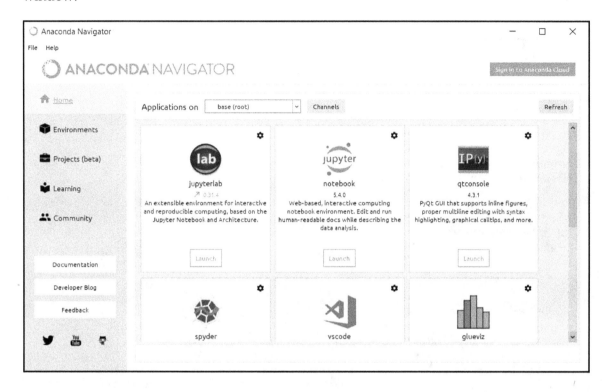

When Jupyter is running, we can get some details on the installation by using the **File** | **About** menu, which will provide a dialog box like this one, which is showing some details on the Jupyter installation, as follows:

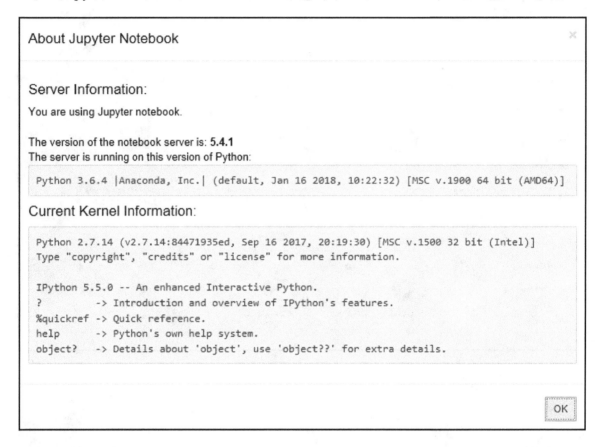

If you start Jupyter from the command line directly, Jupyter will open in a new browser window and you will see some of the logging entries that will display on your Terminal window, noting the progress being made in your use:

```
[I 19:01:01.264 NotebookApp] Serving notebooks from local directory: C:\Users\Dan
[I 19:01:01.264 NotebookApp] 0 active kernels
[I 19:01:01.264 NotebookApp] The Jupyter Notebook is running at:
[I 19:01:01.279 NotebookApp] http://localhost:8888/?token=e03d198ad9a983fae5a7d278ed49869770efee689ed6f9dc
[I 19:01:01.279 NotebookApp] Use Control-C to stop this server and shut down all kernels (twice to skip confirmation).
[C 19:01:01.279 NotebookApp]

    Copy/paste this URL into your browser when you connect for the first time,
    to login with a token:
        http://localhost:8888/?token=e03d198ad9a983fae5a7d278ed49869770efee689ed6f9dc
[I 19:01:02.764 NotebookApp] Accepting one-time-token-authenticated connection from ::1
```

Note that the last line of the log is the instruction you must use to stop the server (press *Ctrl* + *C* in the command-line window where the server is running).

If you press *Ctrl* + *C* in that window, the Jupyter server will shut down gracefully:

```
[W 17:26:36.688 NotebookApp] 404 GET /favicon.ico (::1) 62.00ms
referer=None
[W 17:26:36.750 NotebookApp] 404 GET /favicon.ico (::1) 0.00ms referer=None
[I 17:28:24.891 NotebookApp] Interrupted...
[I 17:28:24.891 NotebookApp] Shutting down kernels
```

You will notice that the Anaconda package has been installed on your application menu for further use:

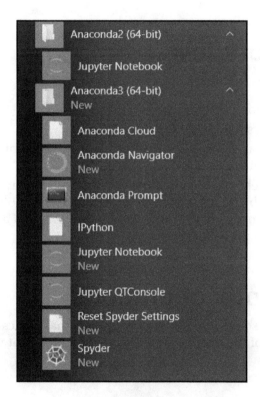

Notebook structure

A Jupyter Notebook is fundamentally a JSON file with a number of annotations. The main parts of the Notebook are as follows:

- **Metadata**: A data dictionary of definitions used to set up and display the Notebook
- **Notebook format**: Version numbers of the software used to create the Notebook (the version number is used for backward compatibility)
- **List of cells**: There are different types of cells for markdown (display), code (to execute), and output (of the code type cells)

Notebook workflow

The typical workflow is as follows:

- Create a new Notebook for a project or data analysis.
- Add your analysis steps, coding, and output.
- Surround your analysis with organizational and presentational markdown to communicate an entire story.
- Interactive Notebooks (that include widgets and display modules) would then be used by others by modifying parameters and data to note the effects of their changes. Your markdown would present the cases that a user may want to investigate, and probable results.

Basic Notebook operations

In this section, we will describe the different operations that you can perform on your Jupyter Notebook. Most of the operations are menu functions that will change your display accordingly.

File operations

Let's walk through the basic file operations.

From the **Files** tab, we can see a list of files and folders in the current Notebook/disk folder. If we select (check) one of the files, we will see the top-left menu change:

We now have choices of **Duplicate**, **Rename**, and delete (the trash icon). Note that the number of files selected, **1**, is displayed in the box as well.

Duplicate

If we hit the **Duplicate** button, we get a confirmation prompt with the name of the file that's been selected for duplication:

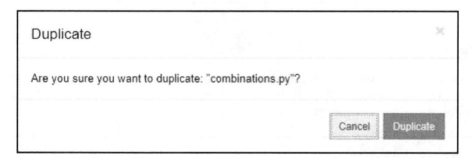

Cancel will close the dialog. **Duplicate** will create another copy of the file with an appended copy number, as shown in the following screenshot. The original filename has been used with the addition of -Copyn in the filename, where n is the copy number. Note the original file extension, .py, has been maintained in the new file:

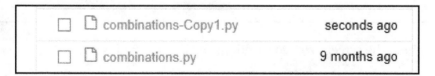

Rename

Similarly, if we hit the **Rename** button, another dialog box will appear to prompt the new filename to apply. The main filename has been highlighted as it assumes you want to maintain the file extension as the file type has not changed:

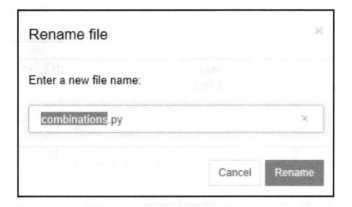

Delete

We can also delete the file by clicking on the trashcan icon. This brings up a confirmation dialog box as follows. I like that they changed the background of **Delete** to red to make sure that you don't just happily click it:

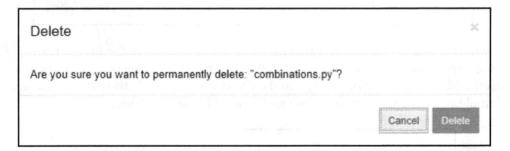

At the top right of the screen, we have options for **Upload** and **New**.

Upload

The **Upload** button is more meaningful when the Notebook is stored on a web server. When running it on your desktop, it allows you to move files easily from one part of your Notebook to another. If you click this button, you are presented with a file selector dialog box. The following screenshot is specific to a Windows environment, but a similar display is presented on macOS. Once you select a file, it will be added to your Notebook space:

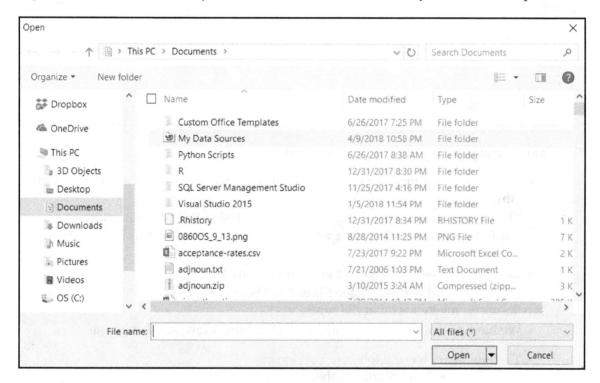

New text file

If we opt to create a **Text File**, we are presented with a new browser panel in the Jupyter text editor (I have shrunk down the size of the screen so that the display fits the boundaries of this book):

There are several points of interest on this screen:

- We are in a new browser panel (the Notebook display is still present in the **Other** tab).
- The name of the new file is `untitled1.txt`. Using the same convention as duplication, the new filename starts with `untitled.txt` and is incremented as needed.
- Curiously, it mentions when the file was created.
- In the top-right corner, we see **Plain Text**. So, we might expect to see some other description here for other file types.
- We have a new menu, which includes **File**, **Edit**, **View**, and **Language**.

- The **File** menu has the following options:
  - **New**: Starts another new text window.
  - **Save**: Save or updates the current text file into the Notebook area.
  - **Rename**: Changes the name of the file (unlikely, as you would want to keep the `untitledn` name that's provided).
  - **Download**: Again, an option that makes more sense if your Notebook is running on the web. As explained for upload, download on a desktop installation allows you to copy a file to another part of your machine.

- The **Edit** menu has the following options:
    - **Find**: Searches for a string.
    - **Find & Replace**: Searches and replaces a string.
    - **Separator**: Below this line is adjusting the text editor in use.
    - **Key Map**: Set your own function mapping for your keyboard.
    - **Default**: Checked as it is the default choice. This means using the default text editor.
    - **Sublime Text**: If you would prefer to use the Sublime editor.
    - **Vim**: If you would prefer to use VIM.
    - **emacs**: If you would prefer to use emacs.

- The **View** menu only has an option to **Toggle Line Numbers**. I imagine future revisions of the package will have additional features. Similarly, for other file types, the menu may change.
- The **Language** menu allows you to specify whether this text file is a specific type of programming file. This allows for syntax highlighting, which is a major feature of source editors. The list is extensive:

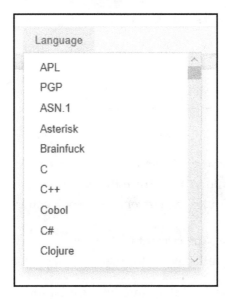

New folder

The **Folder** option creates a new folder with the naming convention `untitled` folder.

New Python 3

The new **Python 3** option creates a new Python 3 Notebook. You are presented with a new browser panel with a similar naming convention, as shown in the following screenshot.

This is a very different presentation, where Python code is expected to be entered in the cells on the page with results displayed in each cell.

There is an extensive menu with **File**, **Edit**, **View**, **Insert**, **Cell**, **Kernel**, and **Help** options. We have a fairly complete **Integrated Development Environment (IDE)** for creating Python coding:

The **File** menu has the following options:

- **New Notebook**: Starts a new Notebook (another browser panel like this one)
- **Open...**: Selects a file to open from the Notebook files view
- **Make a Copy...**: Copies the current Notebook completely into another browser panel
- **Rename...**: Renames the current Notebook
- **Save and Checkpoint**: Saves the current Notebook and records a checkpoint

 A checkpoint is a point in time where all information about a Notebook is preserved. You can have many checkpoints and return the state of your Notebook to the previous checkpoint state at any time. This is an excellent way to give yourself the room to try out a new angle on your analysis without risking losing what you have done so far.

- **Revert to Checkpoint**: Reverts your Notebook to a previous checkpoint
- **Print Preview**: Presents a preview of the printed form of your Notebook
- **Download as**: Downloads the Notebook in a variety of formats:
    - IPython Notebook (its current form)
    - IPython
    - HTML representation
    - Markdown a specialized display format
    - REST – Restructured Text, which is an easy to read, plain text markup
    - PDF
    - Presentation
- **Close and Halt**: Closes the current Notebook and stops any running scripts

 Each of the rectangular work areas in your Notebook is a cell. The innermost text area is where you enter code. Below that (but within the surrounding rectangle), the results of each code stop will be displayed.

- The **Edit** menu has the following options:
    - **Copy Cells**: Copies cells from the clipboard to the current cursor position.
    - **Paste Cells Above**: Pastes cells from the clipboard above the current cell.
    - **Paste Cells Below**: Pastes cells from the clipboard below the current cell.
    - **Paste Cells & Replace**: Pastes the cells from the clipboard on top of the current cell.
    - **Delete Cells**: Deletes the current cells.
    - **Undo Delete Cells**: Reverts the last delete cells invocation.
    - **Split Cell**: Splits up a cell from the current cursor position.
    - **Merge Cell Above**: Merges the current cell with the one above.
    - **Merge Cell Below**: Merges the current cell with the one below.

- **Edit Notebook Metadata**: Every Notebook has underlying metadata which describes the characteristics of the Notebook. Advanced users can manipulate this data directly in order to adjust features more readily. For example, the current Notebook metadata looks like the following screenshot:

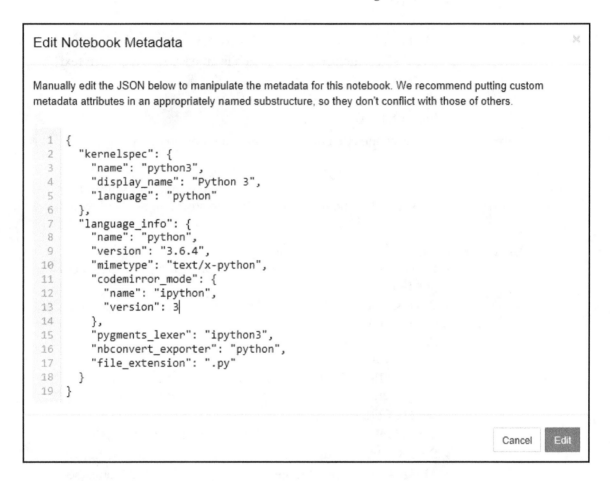

- **Find and Replace**: Allows for find and replace among the selected cells. There is a standardized dialog box for this, as shown in the following screenshot:

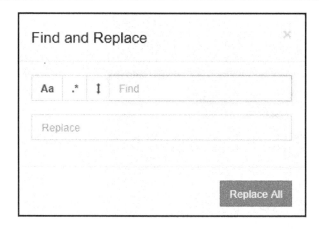

- As seen in the preceding screenshot, the parameters and their functions are as follows:
  - The **Aa** icon toggle determines whether a case-insensitive search is made
  - The * icon toggle determines whether a regex search is made
  - The stacked lines icon toggle determines whether a replace will be made
  - The **Find** text block presents the search criteria
  - The **Replace** text block is used for the replacement text
- The **View** menu has the following options:
  - **Toggle Header**: Toggles the display of the Jupyter logo and filename
  - **Toggle Toolbar**: Toggles the display of the toolbar
  - **Cell Toolbar**: Toggles the display of the cell action icons
- The **Insert** menu has the following options:
  - **Insert Cell Above**: Adds a new cell above the current one
  - **Insert Cell Below**: Adds a new cell below the current one
- The **Cell** menu has the following options:
  - **Run Cells**: Runs the selected (or all) cells.
  - **Run Cells and Select Below**: Runs the current cells down and creates a new one below.
  - **Run Cells and Insert Below**: Runs the current cells and creates a new one above.
  - **Run All**: Runs all cells.
  - **Run All Above**: Runs all cells prior to the current cell.

- **Run All Below**: Runs all cells below the current cell.
- **Cell Type**: Changes the type of cell selected to **Code**, **Markdown**, or **NBConvert**. There is an automatic message that is displayed, noting that all cells are, by default, Code type.
- **Current Outputs** and **All Output** have options to toggle their display.
- The **Kernel** menu has the following options:
  - **Interrupt**: Send a keyboard interrupt, *Ctrl + C*, to the kernel. This is useful if your code is in an endless loop.
  - **Restart**: Restart the kernel.
  - **Restart & Clear Output**: Restart the kernel and clear all output anew.
  - **Restart & Run All**: Restart the kernel and run all cells.
  - **Reconnect**: Connect back to a remote Notebook.
  - **Change kernel**: Not useful as only Python 2 is available at this point.
- The **Help** menu has the following options:
  - **User Interface Tour**: Walks the user through a UI tour
  - **Keyboard Shortcuts**: Presents a list of built-in keyboard shortcuts
  - **Notebook Help**: Presents help topics on the Notebook
  - **Markdown**: Description of the markdown available within a Notebook
  - **Python Reference**, **IPython Reference**, **NumPy Reference**, **SciPy Reference**, **Matplotlib Reference**, **SymPy Reference**, **Pandas Reference**: Help topics on the various languages and packages that can be used in Notebooks
  - **About**: A standard about box

There is an icon panel below the menu that has shortcut icons for the preceding functions:

- **Floppy disk icon**: Save and Checkpoint.
- **Plus sign**: Insert cell below.
- **Scissors**: Cut selected cells.
- **Duplicate pages**: Copy selected cells.
- **Up arrow**: Move selected cells up.
- **Down arrow**: Move selected cells down.
- **An icon that looks like a speaker**: Run the current cell.

- **Black square**: Interrupt the kernel.
- **Circular arrow**: Restart the kernel (with dialog).
- A drop-down menu for display characteristics:
    - **Code**
    - **Markdown**
    - **Raw NBConvert**
    - **Heading**
- **Keyboard**: Open the command palette.
- Change the current toolbar in use. Clicking on the **Cell Toolbar** button auto-displays the **Cell Toolbar** choice from the **View** menu.

Security in Jupyter

Jupyter Notebooks are created in order to be shared with other users, in many cases over the internet. However, Jupyter Notebooks can execute arbitrary code and generate arbitrary code. This can be a problem if malicious aspects have been placed in a Notebook. The default security mechanisms for Jupyter Notebooks include the following:

- Raw HTML is always sanitized (checked for malicious coding). Further information can be found at `https://developers.google.com/caja`.
- You cannot run external JavaScript.
- Cell contents (especially HTML and JavaScript) is not trusted (requires user validation to continue).
- The output from any cell is not trusted.
- All other HTML or JavaScript is never trusted, and clearing the output will cause the Notebook to become trusted when saved.

Security digest

Notebooks can also use a security digest to ensure the correct user is modifying the contents. A digest takes into account the entire contents of the Notebook and a secret (only known by the Notebook creator). This combination ensures that malicious coding is not going to be added to a Notebook.

You can add a security digest to a Notebook by using the following command:

```
~/.jupyter/profile_default/security/notebook_secret
```

Here, you replace the `notebook_secret` part with your secret.

Trust options

You can specifically apply your trust to a Notebook by using the following command-line option:

```
jupyter trust /path/to/notebook.ipynb
```

Or you can do it once the Notebook is opened by the **File** | **Trusted Notebook** menu option.

Configuration options for Jupyter

You can configure some of the display parameters that are used when presenting Notebooks. These are configurable due to the use of a product (**CodeMirror**) to present and modify the Notebook. CodeMirror is a JavaScript-based editor for use within web pages (Notebooks).

The list of configurable options is still in development. Some of the options are as follows:

- **Line-separator**: The character used to separate text lines
- **Theme**: The overall theme of presentation used in the Notebook
- **Indent-unit**: How many spaces to indent blocks of coding

To change the configuration of one of the options, you can open the JavaScript window of your browser, enter the coding to modify an option, and then load your Notebook. Then, the modifications you make will be applied to the Notebook presentation. There is further documentation on this, which is available at `https://codemirror.net/doc/manual.html#option_indentUnit`.

For example, to change the indentation (**Indent-unit**) for your Notebook, you would use the following JavaScript:

```
var mycell = Jupyter.notebook.get_selected_cell();
var cell_config = mycell.config;
var code_patch = {
    CodeCell:{
      cm_config:{indentUnit:2}
    }
  }
cell_config.update(code_patch)
```

You have now seen all of the standard operations that are available to you in a Jupyter Notebook.

Summary

In this chapter, we investigated the various user interface elements that are available in a Notebook. We learned how to install the software on a macOS or a Microsoft PC. We were exposed to the Notebook structure. We saw the typical workflow that's used when developing a Notebook, and we walked through the user interface operations that are available in a Notebook. Lastly, we saw some of the configuration options that are available to advanced users for their Notebook.

In the next chapter, we will learn all about Python scripting in a Jupyter Notebook.

Jupyter Python Scripting 2

Jupyter was originally IPython, an interactive version of Python to be used as a development environment. As such, most of the features of Python are available to you when developing your Notebook.

In this chapter, we will cover the following topics:

- Basic Python scripting
- Python dataset access (from a library)
- Python pandas
- Python graphics
- Python random numbers

Basic Python in Jupyter

We must open a Python section of our Notebook to use Python coding. So, start your Notebook, and then in the upper-right menu, select **Python 3**:

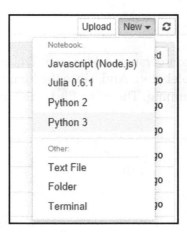

This will open a Python window to work in:

As mentioned in the previous chapter, the new window shows an empty cell so that you can enter Python code.

Let's give the new work area a name, `Learning Jupyter 5, Chapter 2`. Autosave should be on (as you can see next to the title). With an accurate name, we can find this section again easily from the Notebook home page. If you select your browser's **Home** tab and refresh it, you will see this new window name being displayed:

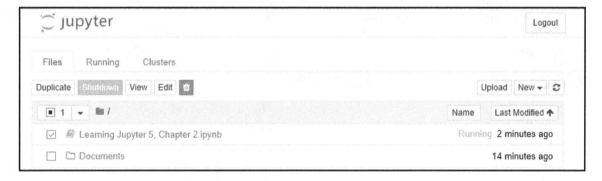

Note that it has an Notebook icon versus a folder icon. The extension that's automatically assigned is `.ipynb` (Python Notebook). And, since the item is in a browser in a Jupyter environment, it is marked as running. There is a file by that name in your directory on the disk as well:

If you open the .ipynb file in a text editor, you will see the basic contents of a Jupyter node (as mentioned in the *Notebook structure* section in Chapter 1, *Introduction to Jupyter*). We have one empty cell and metadata about the Notebook:

```json
{
 "cells": [
  {
   "cell_type": "code",
   "execution_count": null,
   "metadata": {},
   "outputs": [],
   "source": []
  }
 ],
 "metadata": {
  "kernelspec": {
   "display_name": "Python 3",
   "language": "python",
   "name": "python3"
  },
  "language_info": {
   "codemirror_mode": {
    "name": "ipython",
    "version": 3
   },
   "file_extension": ".py",
   "mimetype": "text/x-python",
   "name": "python",
   "nbconvert_exporter": "python",
   "pygments_lexer": "ipython3",
   "version": "3.6.4"
  }
 },
 "nbformat": 4,
 "nbformat_minor": 2
}
```

We can now enter Python coding into the cells. For example:

- Type in some Python in the first cell
- Add another cell to the end (using the **Insert Cell Above** or **Insert Cell Below** menu command):

```python
name = "Dan"
age = 37
```

- And in the second cell, enter the Python code that references the variables from the first cell:

```
print(name + ' is ' + str(age) + ' years old.')
```

- We will then have this display:

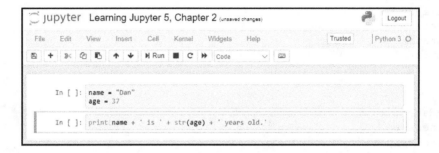

It's important to note that Jupyter color codes your Python (just as a decent editor would), and that we have these empty braces to the left of each code block.

If we execute **Run All**, the results are displayed inline:

```
In [4]:  name = "Dan"
         age = 37

In [5]:  print name + ' is ' + str(age) + ' years old.'

         Dan is 37 years old.
```

We now have the braces filled in with cell numbers, and the output of cells is appended to the bottom of each cell. It's important to note that cell two was able to reference variables that were declared in cell one.

If we either wait for autosave to kick in or hit the save icon (the leftmost icon of a diskette), we will update the `.pynb` file on the disk with our results:

```
{
  "cells": [
    {
      "cell_type": "code",
      "execution_count": null,
      "metadata": {},
      "outputs": [],
```

```
  "source": [
   "name = \"Dan\"\n",
   "age = 37"
  ]
 },
 {
  "cell_type": "code",
  "execution_count": null,
  "metadata": {},
  "outputs": [],
  "source": [
   "print(name + ' is ' + str(age) + ' years old.')"
  ]
 }
],
"metadata... as above
```

It's interesting that Jupyter keeps track of the output last generated in the saved version of the file. You can also clear the output using the **Cell| All Ouput | Clear** command.

If you were then to rerun your cells (using **Cell | Run All**), the output would be regenerated (and saved via autosave). The cell numbering is incremented if you do this – Jupyter is keeping track of the latest version of each cell.

Similarly, if you were to close the browser tab, refresh the display in the **Home** tab, find the new item we created (Learning Jupyter 5, Chapter 2.pynb) and click on it, the new tab (as created previously) will be displayed, showing the outputs that we generated when we last run it.

If you open the server command-line window (where the Jupyter service is running), you will see a list of the actions that we have made during our session:

```
[I 13:41:42.769 NotebookApp] Accepting one-time-token-authenticated connection from ::1
[I 13:45:23.392 NotebookApp] Creating new notebook in
[I 13:45:24.385 NotebookApp] Kernel started: 2b68049e-f8d9-4c3c-9b59-ac89ddf023db
[I 13:45:25.657 NotebookApp] Adapting to protocol v5.1 for kernel 2b68049e-f8d9-4c3c-9b59-ac89ddf023db
[I 13:47:24.315 NotebookApp] Saving file at /Untitled3.ipynb
[I 13:57:24.321 NotebookApp] Saving file at /Learning Jupyter 5, Chapter 2.ipynb
[I 13:57:34.736 NotebookApp] Saving file at /Learning Jupyter 5, Chapter 2.ipynb
[I 13:57:55.903 NotebookApp] Saving file at /Learning Jupyter 5, Chapter 2.ipynb
[I 13:58:20.953 NotebookApp] Saving file at /Learning Jupyter 5, Chapter 2.ipynb
[I 13:59:24.251 NotebookApp] Saving file at /Learning Jupyter 5, Chapter 2.ipynb
[I 14:01:24.313 NotebookApp] Saving file at /Learning Jupyter 5, Chapter 2.ipynb
[I 14:03:10.387 NotebookApp] Saving file at /Learning Jupyter 5, Chapter 2.ipynb
```

The logging entries are at a high level. There may be a way to increase the logging level if there is some difficulty being encountered.

Python data access in Jupyter

Now that we have seen how Python works in Jupyter, including the underlying encoding, how does Python access a large dataset of work in Jupyter?

I started another view for pandas, using Python data access as the name. From here, we will read in a large dataset and compute some standard statistics on the data. We are interested in seeing how we use pandas in Jupyter, how well the script performs, and what information is stored in the metadata (especially if it is a larger dataset).

Our script accesses the iris dataset that is built-in to one of the Python packages. All we are looking to do is read in a slightly large number of items and calculate some basic operations on the dataset. We are really interested to see how much of the data is cached in the .pynb file.

The Python code is as follows:

```python
# import the datasets package
from sklearn import datasets

# pull in the iris data
iris_dataset = datasets.load_iris()
# grab the first two columns of data
X = iris_dataset.data[:, :2]

# calculate some basic statistics
x_count = len(X.flat)
x_min = X[:, 0].min() - .5
x_max = X[:, 0].max() + .5
x_mean = X[:, 0].mean()

# display our results
x_count, x_min, x_max, x_mean
```

I broke these steps into a couple of cells in Jupyter:

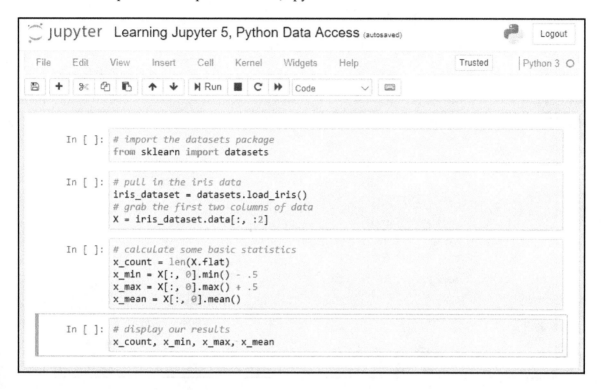

Now, run the cells (using **Cell** | **Run All**) and we will get the following display. The only difference is the last out line where our values are displayed:

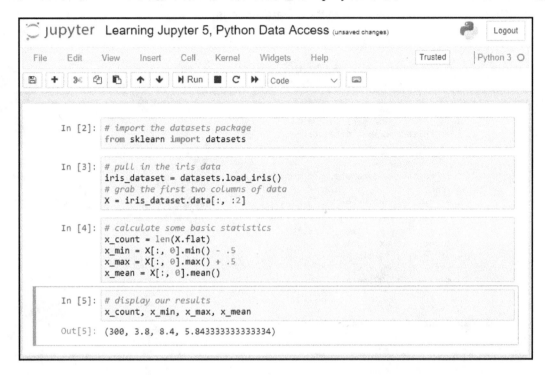

It seemed to take longer to load the library (the first time I ran the script) than to read the data and calculate the statistics.

If we look in the .pynb file for this Notebook, we can see that none of the data is cached in the .pynb file. We simply have code references to the library, our code, and the output from when we last calculated the script:

```
{
 "cells": [
  {
   "cell_type": "code",
   "execution_count": 2,
   "metadata": {},
   "outputs": [],
   "source": [
    "# import the datasets package\n",
    "from sklearn import datasets"
   ]
  },
```

```
  {
   "cell_type": "code",
   "execution_count": 3,
   "metadata": {},
   "outputs": [],
   "source": [
    "# pull in the iris data\n",
    "iris_dataset = datasets.load_iris()\n",
    "# grab the first two columns of data\n",
    "X = iris_dataset.data[:, :2]"
   ]
  },
  {
   "cell_type": "code",
   "execution_count": 4,
   "metadata": {},
   "outputs": [],
   "source": [
    "# calculate some basic statistics\n",
    "x_count = len(X.flat)\n",
    "x_min = X[:, 0].min() - .5\n",
    "x_max = X[:, 0].max() + .5\n",
    "x_mean = X[:, 0].mean()"
   ]
  },
  {
   "cell_type": "code",
   "execution_count": 5,
   "metadata": {},
   "outputs": [
    {
     "data": {
      "text/plain": [
       "(300, 3.8, 8.4, 5.843333333333334)"
      ]
     },
     "execution_count": 5,
     "metadata": {},
     "output_type": "execute_result"
    }
   ],
   "source": [
    "# display our results\n",
    "x_count, x_min, x_max, x_mean"
   ]
  }
]...
```

Python pandas in Jupyter

One of the most widely used features of Python is pandas. The pandas are built-in libraries of data analysis packages that can be used freely. In this example, we will develop a Python script that uses pandas to see if there is any affect of using them in Jupyter.

I am using the Titanic dataset from `https://www.kaggle.com/c/titanic/data`. I am sure that the same data is available from a variety of sources.

 Note that you have to sign up for **Kaggle** in order to download the data. It's free.

Here is our Python script that we want to run in Jupyter:

```
from pandas import *
training_set = read_csv('train.csv')
training_set.head()
male = training_set[training_set.Sex == 'male']
female = training_set[training_set.Sex =='female']
womens_survival_rate = float(sum(female.Survived))/len(female)
mens_survival_rate = float(sum(male.Survived))/len(male)
womens_survival_rate, mens_survival_rate
```

The result is that we calculate the survival rates of the passengers based on sex.

We create a new Notebook, enter the script into the appropriate cells, include adding displays of calculated data at each point, and produce our results.

Here is our Notebook laid out, where we added displays of calculated data at each cell:

When I ran this script, I had two problems:

On Windows, it is common to use a backslash (\) to separate parts of a filename. However, this coding uses the backslash as a special character. So, I had to change over to using a forward slash (/) in my .csv file path.

The dataset column names are taken directly from the file and are case-sensitive. In this case, I was originally using the sex field in my script, but in the .csv file, the column is named Sex. Similarly, I had to change survived to Survived.

The final script and results look like this when we run it:

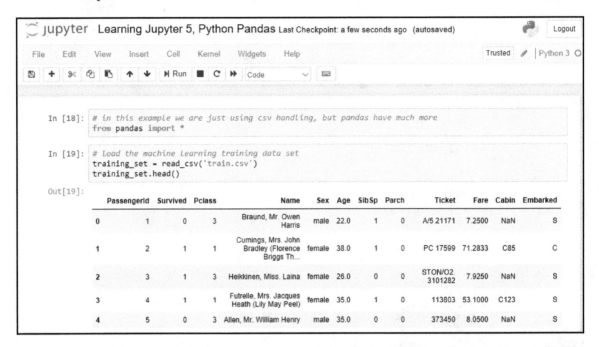

I have used the `head()` function to display the first few lines of the dataset. It is interesting the amount of detail that is available for all of the passengers.

If you scroll down, you will see the results:

```
In [20]:  # split the data by sex
          male = training_set[training_set.Sex == 'male']
          female = training_set[training_set.Sex =='female']

In [21]:  # calculate survival rates based on sex
          womens_survival_rate = float(sum(female.Survived))/len(female)
          mens_survival_rate = float(sum(male.Survived))/len(male)
          womens_survival_rate, mens_survival_rate

Out[21]:  (0.7420382165605095, 0.18890814558058924)
```

We can see that 74% of the survivors were women versus just 19% men. I would like to think that chivalry is not dead.

It's curious that the results do not total to 100%. However, like every other dataset I have seen, there is missing and/or inaccurate data present.

Python graphics in Jupyter

How do Python graphics work in Jupyter?

I started another view for this named Python graphics so as to distinguish the work.

If we were to build a sample dataset of baby names and the number of births in a year of that name, we could then plot the data.

The Python coding is simple:

```
import pandas
import matplotlib

%matplotlib inline

# define our two columns of data
baby_name = ['Alice','Charles','Diane','Edward']
number_births = [96, 155, 66, 272]

# create a dataset from the to sets
dataset = list(zip(baby_name,number_births))
dataset

# create a Python dataframe from the dataset
df = pandas.DataFrame(data = dataset, columns=['Name', 'Number'])
df

# plot the data
df['Number'].plot()
```

The steps for the script are as follows:

1. Import the graphics library (and data library) we need
2. Define our data
3. Convert the data into a format that allows for an easy graphical display
4. Plot the data

We would expect a resultant graph of the number of births by baby name.

Taking the previous script and placing it into cells of our Jupyter node, we get something that looks like this:

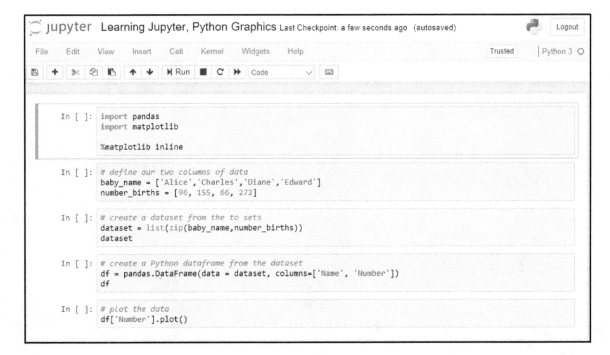

- I have broken the script into different cells for easier readability. Having different cells also allows you to develop the script easily, step-by-step, where you can display the values computed so far to validate your results. I have done this in most of the cells by displaying the dataset and dataframe at the bottom of those cells.
- When we run this script (**Cell | Run All**), we can see the results at each step being displayed as the script progresses:

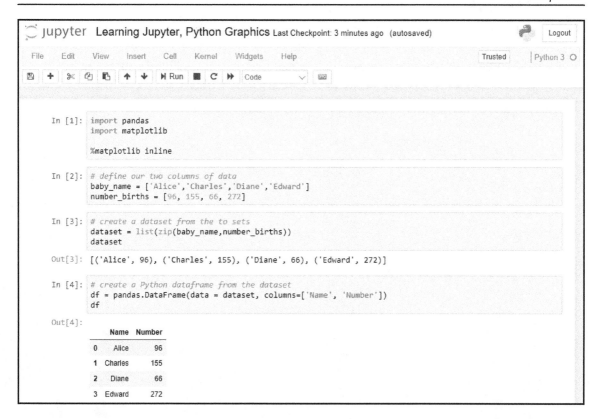

- And finally, we can see our plot of the births:

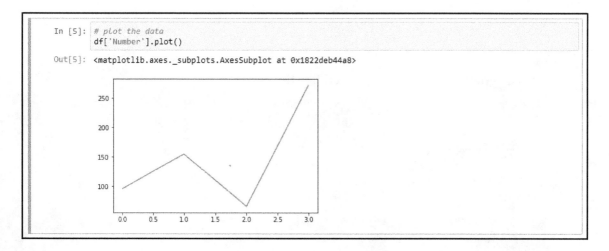

- I was curious about what metadata was stored for this script. Looking into the .ipynb file, you can see the expected value for the formula cells.

- The tabular data display of the dataframe is stored as HTML convenient:

```
...{
    "cell_type": "code",
    "execution_count": 4,
    "metadata": {},
    "outputs": [
     {
      "data": {
       "text/html": [
        "<div>\n",
        "<style scoped>\n",
        "    .dataframe tbody tr th:only-of-type {\n",
        "        vertical-align: middle;\n",
        "    }\n",
        "\n",
        "    .dataframe tbody tr th {\n",
        "        vertical-align: top;\n",
        "    }\n",
        "\n",
        "    .dataframe thead th {\n",
        "        text-align: right;\n",
        "    }\n",
        "</style>\n",
        "<table border=\"1\" class=\"dataframe\">\n",
        "  <thead>\n",
        "    <tr style=\"text-align: right;\">\n",
        "      <th></th>\n",
        "      <th>Name</th>\n",
        "      <th>Number</th>\n",
        "    </tr>\n",...
```

- The graphic output cell is stored like this:

```
{
    {
    "cell_type": "code",
    "execution_count": 5,
    "metadata": {},
    "outputs": [
     {
      "data": {
       "text/plain": [
        "<matplotlib.axes._subplots.AxesSubplot at 0x1822deb44a8>"
       ]
```

```
      },
      "execution_count": 5,
      "metadata": {},
      "output_type": "execute_result"
      },
      {
      "data": {
       "image/png": "iVBORw0...
  "<a hundred lines of hexcodes>
  ...VTRitYII=\n",
      "text/plain": [
       "<matplotlib.figure.Figure at 0x1822e26a828>"
      ]
      },...
```

Where the `image/png` tag contains a large hex digit string representation of the graphical image displayed on screen (I abbreviated the display in the coding that's shown). So, the actual generated image is stored in the metadata for the page.

So, rather than a cache, Jupyter is remembering the output from when each cell was last executed.

Python random numbers in Jupyter

For many analyses, we are interested in calculating repeatable results. However, much of this analysis relies on some random numbers being used. In Python, you can set `seed` for the random number generator to achieve repeatable results with the `random.seed()` function.

In this example, we simulate rolling a pair of dice and look at the outcome. We would expect the average total of the two dice to be six, which is the half-way point between the faces.

The script we are using is as follows:

```
# using pylab statistics and histogram
import pylab
import random

# set random seed so we can reproduce results
random.seed(113)
samples = 1000

# declare our dataset store
dice = []
```

```
# generate and save the samples
for i in range(samples):
    total = random.randint(1,6) + random.randint(1,6)
    dice.append(total)

# compute some statistics on the dice throws
print("Throw two dice", samples, "times.")
print("Mean of", pylab.mean(dice))
print("Median of", pylab.median(dice))
print("Std Dev of", pylab.std(dice))

# display a histogram of the results
pylab.hist(dice, bins= pylab.arange(1.5,12.6,1.0))
pylab.xlabel("Pips")
pylab.ylabel("Count")
pylab.show()
```

Once we have the script in Jupyter and execute it, we will get the following result:

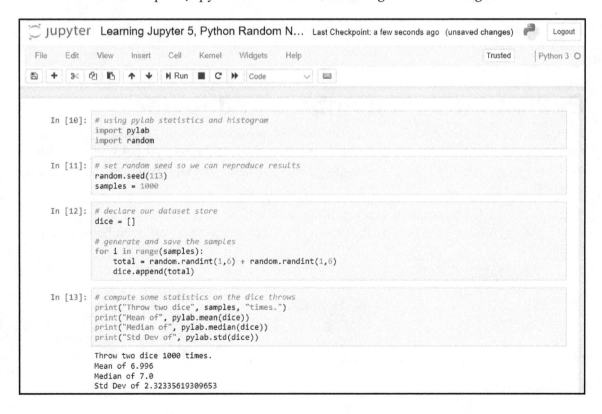

I added some more statistics. I'm not sure that I would have counted on such a high standard deviation. If we increased the number of `samples`, this would decrease.

The resulting graph was opened in a new window, much as it would if you ran this script in another Python development environment:

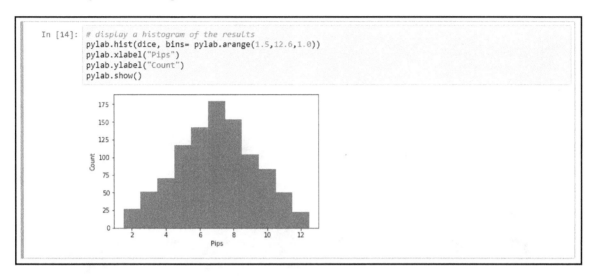

```
In [14]:  # display a histogram of the results
          pylab.hist(dice, bins= pylab.arange(1.5,12.6,1.0))
          pylab.xlabel("Pips")
          pylab.ylabel("Count")
          pylab.show()
```

The graphic looks a little more jagged than I would have expected for a thousand samples.

Summary

In this chapter, we walked through a simple Notebook and the underlying structure. Then, we saw an example of using pandas and looked at a graphics example. Finally, we looked at an example using random numbers in a Python script.

In the next chapter, we will learn about R scripting in a Jupyter Notebook.

Jupyter R Scripting 3

Jupyter's native language is Python. Once Jupyter (which was essentially IPython before being renamed) became popular for data analysis, a number of people were interested in using the suite of R programming analysis tools that are available in Jupyter Notebook.

In this chapter, we will cover the following topics:

- Adding R scripting to your installation
- Basic R scripting
- R dataset access (from a library)
- R graphics
- R cluster analysis
- R forecasting
- R predictions

Adding R scripting to your installation

Two big installation platforms are macOS and Windows. There are separate but similar steps for making R scripting available in your Jupyter installation.

Adding R scripts to Jupyter on macOS

If you are operating on macOS, you can add R Scripting using the following command:

```
conda install -c r r-essentials
```

This will start off with a large installation of the R environment, which contains a number of common packages:

```
bos-mpdc7:~ dtoomey$ conda install -c r r-essentials
Fetching package metadata: ......
```

```
Solving package specifications: ........

Package plan for installation in environment /Users/dtoomey/miniconda3:
```

The following packages will be downloaded:

```
    package                    |             build
    ---------------------------|-----------------
    jbig-2.1                   |               0          31 KB
    jpeg-8d                    |               2         210 KB
    libgcc-4.8.5               |               1         785 KB
... <<many packages>>
    r-caret-6.0_62             |        r3.2.2_0a         4.3 MB
    r-essentials-1.1           |        r3.2.2_1a         726 B
    ---------------------------------------------------------
                                          Total:        101.0 MB
```

The following new packages will be installed:

```
    jbig:           2.1-0
    jpeg:           8d-2
... <<many packages>>
    r-xts:          0.9_7-r3.2.2_0a
    r-yaml:         2.1.13-r3.2.2_1a
    r-zoo:          1.7_12-r3.2.2_1a
    zeromq:         4.1.3-0

Proceed ([y]/n)? y

Fetching packages ...
jbig-2.1-0.tar 100% |##############################| Time: 0:00:00    1.59
MB/s
jpeg-8d-2.tar. 100% |##############################| Time: 0:00:00    2.69
MB/s
... <<many packages>>
r-caret-6.0_62 100% |##############################| Time: 0:00:00   11.16
MB/s
r-essentials-1 100% |##############################| Time: 0:00:00  537.43
kB/s
Extracting packages ...
[      COMPLETE       ]|###################################################|
100%
Linking packages ...
[      COMPLETE       ]|###################################################|
100%
```

From there, you will invoke your `notebook` as you normally would:

```
ipython notebook
```

Adding R scripts to Jupyter on Windows

The default installation for Anaconda does not include R—I'm not sure why. Once Anaconda is installed, you need to specially install it using the command line. Then, we will take the plunge and add R scripting, which is very similar to what we did for macOS:

```
conda install -c r notebook r-irkernel
```

This produces a detailed view of the packages that are updated. In my case, it installed a full set of R packages and runtimes, even though I had used R elsewhere on the machine earlier:

```
C:\Users\Dan>conda install -c r notebook r-irkernel
Solving environment: done
==> WARNING: A newer version of conda exists. <==
current version: 4.4.10
latest version: 4.5.4
Please update conda by running
$ conda update -n base conda
## Package Plan ##
environment location: C:\Users\Dan\Anaconda3
added / updated specs:
- notebook
- r-irkernel
The following packages will be downloaded:
package | build
-------------------|-------------------r-pbdzmq-0.2_6 | mro343h889e2dd_0 4.2
MB r
libxml2-2.9.8 | vc14_0 3.2 MB conda-forge
...
r-stringr-1.2.0 | mro343h889e2dd_0 143 KB r
r-repr-0.12.0 | mro343h889e2dd_0 68 KB r
r-irdisplay-0.4.4 | mro343h889e2dd_0 29 KB r
jpeg-9b | vc14_2 314 KB conda-forge
r-r6-2.2.2 | mro343_0 5.4 MB r
r-digest-0.6.13 | mro343h889e2dd_0 168 KB r
-------------------------------------------------------------
Total: 140.2 MB
The following NEW packages will be INSTALLED:
mro-base: 3.4.3-0 r
r-crayon: 1.3.4-mro343h889e2dd_0 r
r-digest: 0.6.13-mro343h889e2dd_0 r
```

```
r-evaluate: 0.10.1-mro343h889e2dd_0  r
...
r-stringr: 1.2.0-mro343h889e2dd_0  r
r-uuid: 0.1_2-mro343h889e2dd_0  r
The following packages will be UPDATED:
_r-mutex: 1.0.0-anacondar_1 r --> 1.0.0-mro_2 r
jpeg: 9b-hb83a4c4_2 --> 9b-vc14_2 conda-forge [vc14]
libxml2: 2.9.7-h79bbb47_0 --> 2.9.8-vc14_0 conda-forge [vc14]
sqlite: 3.22.0-h9d3ae62_0 --> 3.22.0-vc14_0 conda-forge [vc14]
r-base: 3.4.3-h6bb4b03_0 r --> 3.4.1-1 conda-forge
Proceed ([y]/n)? y
Downloading and Extracting Packages
r-pbdzmq 0.2_6:  ###################################### | 100%
libxml2 2.9.8:   ###################################### | 100%
r-irkernel 0.8.11: #################################### | 100%
r-magrittr 1.5:  ###################################### | 100%
r-evaluate 0.10.1: #################################### | 100%
_r-mutex 1.0.0:  ###################################### | 100%
...
r-digest 0.6.13: ############################################################
| 100%
Preparing transaction: done
Verifying transaction: done
Executing transaction: done
```

Now when you start Jupyter and pull down the kernel menu, you will see **R** as a choice:

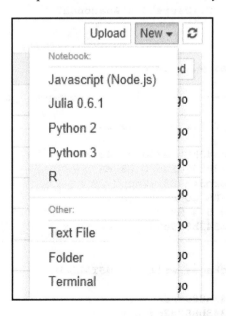

Adding R packages to Jupyter

The standard installation for R under Jupyter has many packages that are commonly used in R programming. However, if you do need to add another package, a small number of steps need to be followed:

1. Close down your Notebook (including the server).
2. In the command-line window, type the following:

```
R install.packages("name of the R package you want to add")
quit()
# answer Yes to save
```

3. Restart your Notebook and the package should be available in your R script, for example, `library (name of the R package you want to add)`.

 Note that you may still have problems in R where the core version of R that you have installed is out of date, so you will need to upgrade it to use a particular library.

R limitations in Jupyter

In this chapter, we used a variety of packages, both pre-installed and installed especially for the example. I have exercised a variety of materials that are available in R under Jupyter and have not found any limitations; you can perform most of the steps in Jupyter that you would have done under the standard R implementations. The only limitation appears to be when you are using *Shiny* or if you are attempting to use extensive markdown:

- For Shiny, I think you are mixing purposes—Jupyter is providing a web experience and so does Shiny—so I'm not sure how to even decide if this should work. This issue is being addressed by the Jupyter development group.
- Using extensive markdown does not appear to be a good idea, either. The intent of markdown was to allow Notebook developers to augment the standard output (of R) in a more illustrative manner. I think that if you are adding extensive markdown to your Notebook, you really need to develop a website—maybe using Shiny and then you will have all of the HTML markdown available.

Basic R in Jupyter

Start a new R Notebook and call it `R Basics`. We can enter a small script just so we can see how the steps progress for R script. Enter the following into separate cells of your Notebook:

myString <- "Hello, World!"
print (myString)

From here, you will end up with a starting screen that looks like this:

We should note the aspects of the R Notebook view:

- We have the R logo in the upper-right corner. You will see this logo running in other R installations.
- There is also the peculiar **R O** just below the R icon. If the **O** unfilled circle displays, the unfilled circle indicates that the kernel is at rest, and the filled circle indicates that the kernel is working.
- The rest of the menu items are the same as the ones we saw previously.

This is a very simple script–set a variable in one cell and then print out its value in another cell. Once executed (**Cell | Run All**), you will see your results:

So, just as if you run the script in an R interpreter, you get your output (with the numerical prefix). Jupyter has counted the statements so that we have incremental numbering of the cells. Jupyter has not done anything special to print out variables for debugging; you will have to do that separately.

If we look at the R server-logging statements (a command-line window was created when we started Jupyter), we will be able to see the actions that took place:

```
$ jupyter notebook
[I 11:00:06.965 NotebookApp] Serving notebooks from local directory:
/Users/dtoomey/miniconda3/bin
[I 11:00:06.965 NotebookApp] 0 active kernels
[I 11:00:06.965 NotebookApp] The Jupyter Notebook is running at:
http://localhost:8888/
[I 11:00:06.965 NotebookApp] Use Control-C to stop this server and shut
down all kernels (twice to skip confirmation).
[I 11:00:17.447 NotebookApp] Creating new notebook in
[I 11:00:18.199 NotebookApp] Kernel started:
518308da-460a-4eb9-9959-1411e31dec69
[1] "Got unhandled msg_type:" "comm_open"
[I 11:02:18.160 NotebookApp] Saving file at /Untitled.ipynb
[I 11:08:27.340 NotebookApp] Saving file at /R Basics.ipynb
[1] "Got unhandled msg_type:" "comm_open"
[I 11:14:45.204 NotebookApp] Saving file at /R Basics.ipynb
```

We started the server, created a new Notebook, and saved it as R Basics. If we open the IPYNB file on disk (using a text editor), we will be able to see the following:

```
{
  "cells": [
    ...<similar to previously displayed>
  ],
  "metadata": {
    "kernelspec": {
```

```
        "display_name": "R",
        "language": "R",
        "name": "ir"
      },
    "language_info": {
      "codemirror_mode": "r",
      "file_extension": ".r",
      "mimetype": "text/x-r-source",
      "name": "R",
      "pygments_lexer": "r",
      "version": "3.4.3"
    }
  },
  ...<omitted>
}
```

This is a little different than what we saw in the prior chapter on Python Notebook coding. In particular, the metadata clearly targets the script cells to be R script. Note that the actual cells are not specific to a language – they are just scripts that will be executed as per the metadata directives.

R dataset access

For this example, we will use the Iris dataset. Iris is built into R installations and is available directly. Let's just pull in the data, gather some simple statistics, and plot the data. This will show R accessing a dataset in Jupyter, using an R built-in package, as well as some available statistics (since we have R), and the interaction with R graphics.

The script we will use is as follows:

```
data(iris)
summary(iris)
plot(iris)
```

If we enter this small script into a new R Notebook, we get an initial display that looks like the following:

I would expect the standard R statistical summary as output, and I know that the Iris plot is pretty interesting. We can see exactly what happened in the following screenshot:

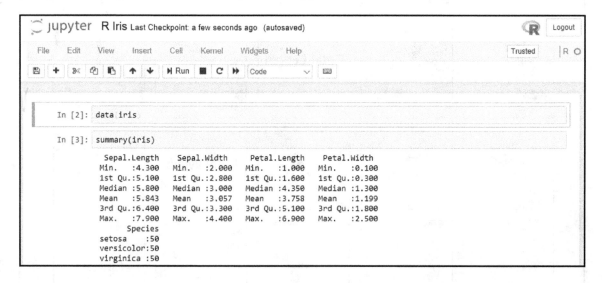

The plot continues in the following screenshot, as it wouldn't fit into a single page:

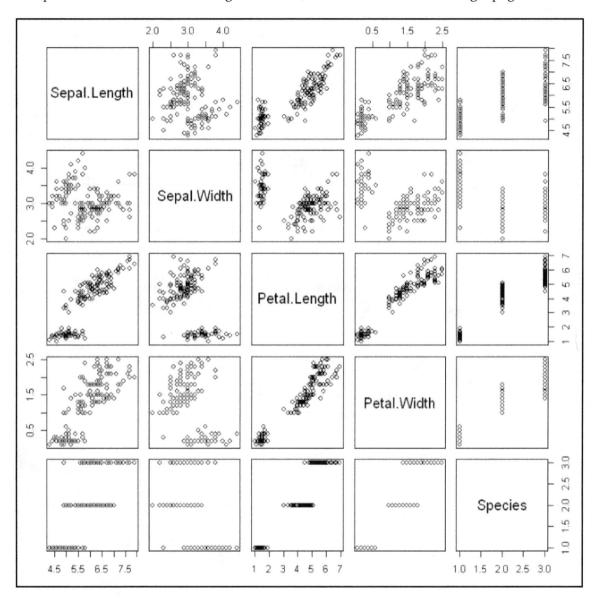

A feature of Jupyter is to place larger plots, such as this, into a viewport that only shows a part of the image. I was able to drag the image out of the viewport window in its entirety for this shot. You can eliminate the viewport boundaries and have the entire output displayed by clicking in the viewport and dragging it.

R visualizations in Jupyter

A common use of R is to use several visualizations, which are available depending on the underlying data. In this section, we will go over some of them to see how R interacts with Jupyter.

R 3D graphics in Jupyter

One of the packages available for 3D graphics is `persp`. The `persp` package draws perspective plots over a 2D space.

We can enter a basic `persp` command in a new Notebook just by using the following command:

```
example(persp)
```

So, we will have something like this in a Notebook:

Once we run the step (**Cell | Run All**), we will see the display that's shown in the following screenshot. The first part is the script involved in generating the graphic (this is part of the example code):

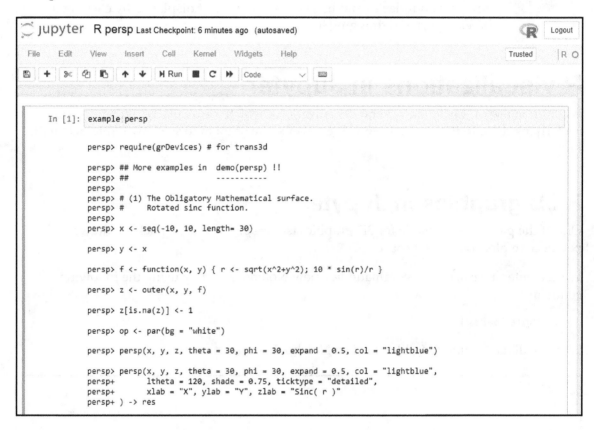

Then, we will see the following graphic display:

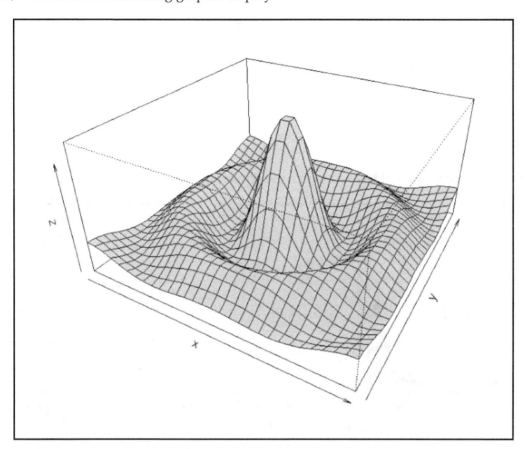

R 3D scatterplot in Jupyter

The R `lattice` package has a Cloud function that will produce 3D scatterplots.

The script we will use is as follows:

```
# make sure lattice package is installed
install.packages("lattice")

# in a standalone R script you would have a command to download the lattice
library - this is not needed in Jupyter

library("lattice")
```

```
# use the automobile data from ics.edu
mydata <-
read.table("http://archive.ics.uci.edu/ml/machine-learning-databases/auto-m
pg/auto-mpg.data")

# define more meaningful column names for the display
colnames(mydata) <- c("mpg", "cylinders", "displacement", "horsepower",
"weight", "acceleration", "model.year", "origin", "car.name")

# 3-D plot with number of cylinders on x axis, weight of the vehicle on the
y axis and miles per gallon on the z axis.
cloud(mpg~cylinders*weight, data=mydata)
```

Prior to running it, we will have something such as this:

Notice that we are using markup type cells for comments about the script steps. They are also denoted without a script line number in the left-hand column.

If you are copying R script into a Jupyter window, you may run across an issue where the print copy you are using has non-standard double quote characters (quotes on the left lean to the left, while quotes on the right lean to the right). Once copied into Jupyter, you will need to change this to normal double quotes (they don't lean; instead, they are vertical).

After running this, we will see the following display:

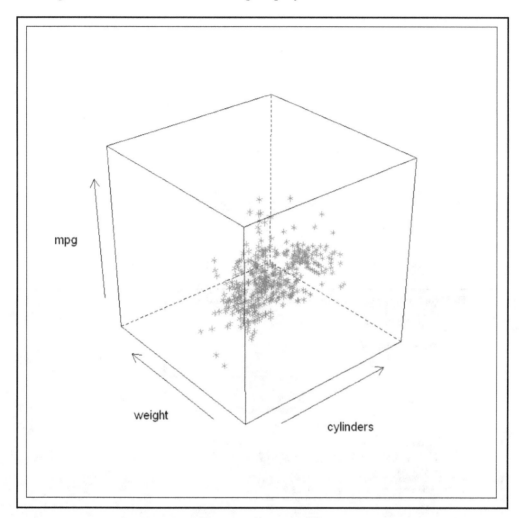

R cluster analysis

In this example, we will use R's cluster analysis functions to determine the clustering in the wheat dataset from `https://uci.edu/`.

The R script we want to use in Jupyter is as follows:

```
# load the wheat data set from uci.edu
wheat <-
read.csv("http://archive.ics.uci.edu/ml/machine-learning-databases/00236/se
eds_dataset.txt", sep="\t")

# define useful column names
colnames(wheat) <-c("area", "perimeter", "compactness", "length", "width",
"asymmetry", "groove", "undefined")

# exclude incomplete cases from the data
wheat <- wheat[complete.cases(wheat),]

# calculate the clusters
set.seed(117) #to make reproducible results
fit <- kmeans(wheat, 5)
fit
```

Once entered into a Notebook, we will have something such as this:

The resulting, generated cluster information is k-means clustering with five clusters of sizes; 39, 53, 47, 29, and 30 (Note that I set the seed value for random number use, so your results will not vary):

```
K-means clustering with 5 clusters of sizes 39, 53, 47, 29, 30

Cluster means:
      area perimeter compactness   length    width asymmetry    groove undefined
1 11.93974  13.28436   0.8489436 5.243872 2.866000  5.585949 5.132795  2.948718
2 14.21057  14.24396   0.8792887 5.488245 3.230075  2.472457 5.050660  1.000000
3 19.15745  16.47553   0.8866957 6.271340 3.772234  3.473553 6.125702  2.000000
4 16.45345  15.35310   0.8768000 5.882655 3.462517  3.913207 5.707655  1.724138
5 11.90700  13.23733   0.8535767 5.216200 2.861767  3.609700 5.073200  2.800000

Clustering vector:
  1   2   3   4   5   6   9  10  11  12  13  14  15  16  17  18  19  20  21  22
  2   2   2   2   2   2   4   4   4   2   2   2   2   2   2   2   2   5   2   2
 23  24  25  26  27  28  29  30  31  32  33  34  35  38  39  40  41  42  43  44
  2   2   2   2   2   2   2   2   2   4   2   2   2   4   4   2   1   2   2   2
 45  46  47  48  49  50  51  52  53  54  55  56  57  58  59  60  61  64  65  66
  4   2   2   2   2   2   2   2   4   2   2   2   2   2   2   2   2   5   5   2
 67  68  69  70  73  74  75  76  77  78  79  80  81  82  83  84  85  86  87  88
  2   2   2   2   2   4   4   4   3   4   4   4   3   3   4   4   3   3   3   3
 89  90  91  92  93  94  95  96  97  98  99 100 101 102 103 104 105 106 107 108
  3   3   3   3   3   3   3   3   3   4   3   3   3   3   4   3   3   3   3   3
109 112 113 114 115 116 117 118 119 120 121 122 123 124 125 126 127 128 129 130
  3   4   3   3   3   3   3   3   3   3   3   3   3   3   3   4   3   4   3   3
131 132 133 134 135 136 137 138 139 142 143 144 145 146 147 148 149 150 151 152
  3   3   3   4   3   3   4   4   4   4   4   4   4   1   1   1   1   5   1   5
153 154 155 156 157 158 159 160 161 162 163 164 165 166 167 168 169 170 171 172
  5   5   1   1   1   1   5   5   1   5   1   5   1   5   5   1   5   1   5   1
173 174 179 182 183 184 185 186 187 188 189 190 191 192 193 194 195 196 197 198
  5   5   1   1   1   1   1   1   1   5   1   1   1   1   1   5   1   5   1   1
199 200 201 202 203 204 205 206 207 208 209 212 215 216 217 218 219 220
  1   5   5   1   5   1   1   1   5   5   1   5   5   5   5   1   5   1

Within cluster sum of squares by cluster:
[1]   71.99806 115.91477 117.61754  54.16095  41.33632
 (between_SS / total_SS =  85.1 %)

Available components:

[1] "cluster"      "centers"      "totss"        "withinss"     "tot.withinss"
[6] "betweenss"    "size"         "iter"         "ifault"
```

So, we generated the information of five clusters (the parameter passed into the fit statement). It is a little bothersome that the cluster sum of squares vary greatly.

R forecasting

For this example, we will forecast the Fraser River levels, given the data from `https://datamarket.com/data/set/22nm/fraser-river-at-hope-1913-1990#!ds=22nm&display=line`. I was not able to find a suitable source, so I extracted the data by hand from the site into a local file.

We will be using the R `forecast` package. You have to add this package to your setup (as described at the start of this chapter).

The R script we will be using is as follows:

```
library(forecast)
fraser <- scan("fraser.txt")
plot(fraser)
fraser.ts <- ts(fraser, frequency=12, start=c(1913,3))
fraser.stl = stl(fraser.ts, s.window="periodic")
monthplot(fraser.stl)
seasonplot(fraser.ts)
```

The output of interest in this example are the three plots: simple plot, monthly, and computed seasonal.

When this is entered into a Notebook, we will get a familiar layout:

```
In [ ]:   install.packages("forecast")
          library(forecast)

In [ ]:   fraser <- scan("fraser.txt")

In [ ]:   plot(fraser)

In [ ]:   fraser.ts <- ts(fraser, frequency=12, start=c(1913,3))

In [ ]:   fraser.stl = stl(fraser.ts, s.window="periodic")

In [ ]:   monthplot(fraser.stl)

In [ ]:   seasonplot(fraser.ts)
```

The simple plot (using the R `plot` command) is like the one that's shown in the following screenshot. There is no apparent organization or structure:

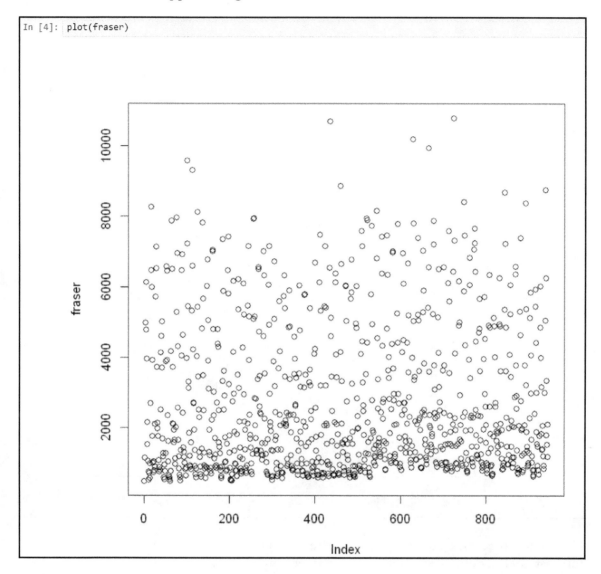

```
In [4]: plot(fraser)
```

The monthly plot (using the `monthplot` command) is like what's shown in the following screenshot. River flows appear to be very consistent within a month:

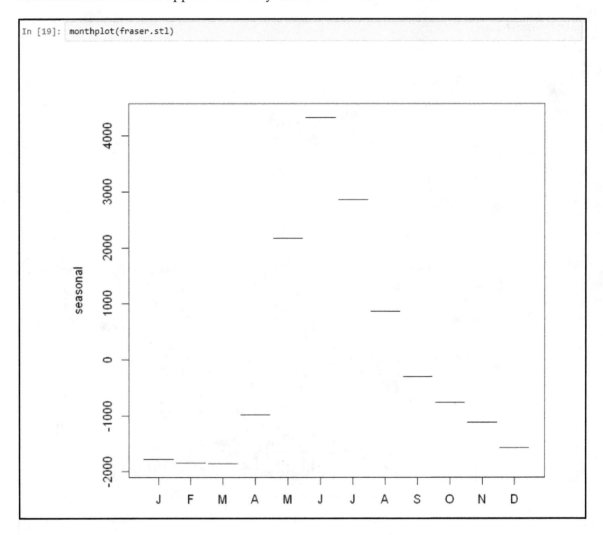

Finally, the `seasonalplot` shows, quite dramatically, what we have been trying to forecast, that is, definite seasonality to the river flows:

```
In [20]: seasonplot fraser.ts
```

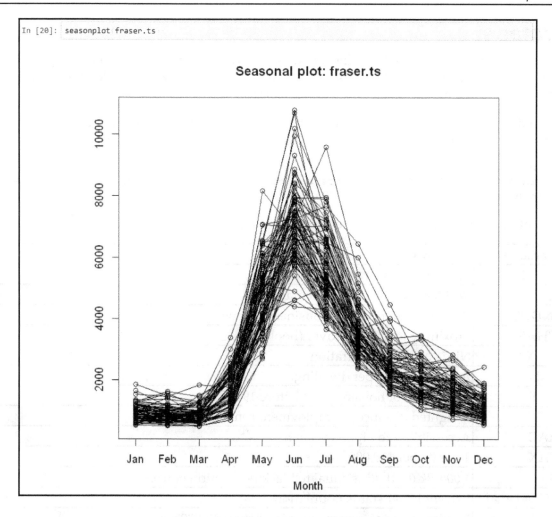

R machine learning

In this section, we will use an approach for machine learning where we will do the following:

- Partition the dataset into a training and testing set
- Generate a model of the data
- Test the efficiency of our model

Dataset

Machine learning works by featuring a dataset that we will break up into a training section and a testing section. We will use the training data to come up with a model. We can then prove or test that model against the testing dataset.

For a dataset to be usable, we need at least a few hundred observations. I am using the housing data from `http://uci.edu`. Let's load the dataset by using the following command:

```
housing <-
read.table("http://archive.ics.uci.edu/ml/machine-learning-databases/housin
g/housing.data")
```

The site documents the names of the variables as follows:

Variables	Description
CRIM	Per capita crime rate
ZN	Residential zone rate percentage
INDUS	Proportion of non-retail business in town
CHAS	Proximity to Charles River (Boolean)
NOX	Nitric oxide concentration
RM	Average rooms per dwelling
AGE	Proportion of housing built before 1940
DIS	Weighted distance to employment center
RAD	Accessibility to highway
TAX	Tax rate per $10,000
B	$1,000(Bk-0.63)^2$ Bk is equal to black population percentage
LSTAT	Percent lower status population
MEDV	Median value of owner-occupied homes $1,000's

So, let's apply these so that we can make sense of the data:

```
colnames(housing) <-
c("CRIM","ZN","INDUS","CHAS","NOX","RM","AGE","DIS","RAD", "TAX",
"PRATIO","B","LSTAT", "MDEV")
```

Now, we can get a summary so that we can get a feel for the values:

```
summary(housing)
```

This results in a screen that looks like this when executed:

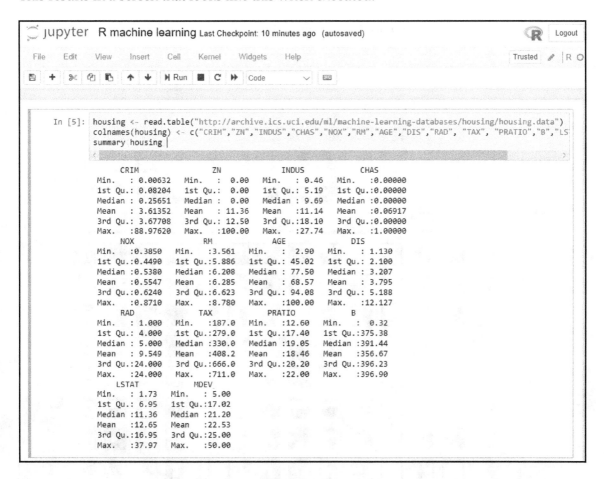

So, the data is somewhat dated. The median values are very low compared to current housing prices in the area.

Also, the statistics are not politically correct – the B factor is a measure of the Black population.

Overall, we have a good number of variables. Which ones are likely candidates for our model? I have found that the best tool for this is a simple regression plot of every variable against every other variable. We can use the following command for this:

```
plot(housing)
```

R does what we want, as shown in the following display:

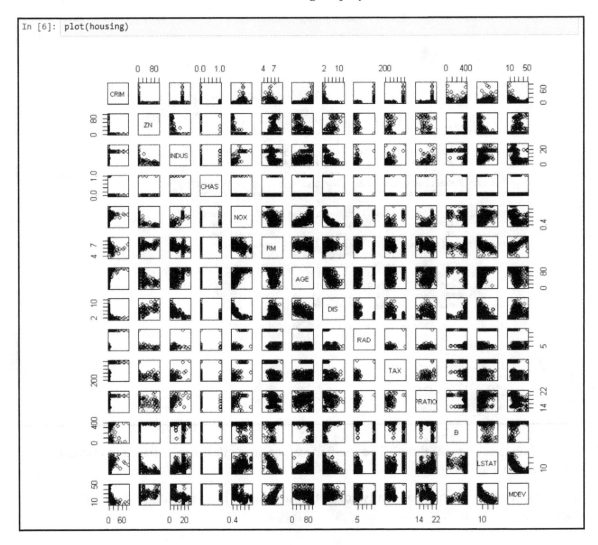

We are looking for either a positive or negative correlation, that is, a somewhat 45-degree line or a negative 45-degree line. Anything that shows a hodgepodge or vertical or horizontal pattern is not going to tell us anything about the data.

The best correlations occurred with RAD (access to highway) and TAX (rate per $1,000). Luckily, most of the variables show a good correlation.

As housing value is our yardstick, let's order the data by it before we partition it, by using the following command:

```
housing <- housing[order(housing$MDEV),]
```

We will be using the caret package to partition the data, so let's load that in:

```
install.packages("caret")
library("caret")
```

Now, we can partition the data:

```
# set the random seed so we can reproduce results
set.seed(311)

# take 3/4 of the data for training
trainingIndices <- createDataPartition(housing$MDEV, p=0.75, list=FALSE)

# split the data
housingTraining <- housing[trainingIndices,]
housingTesting <- housing[-trainingIndices,]

# make sure the paritioning is working
nrow(housingTraining)
nrow(housingTesting)
381
125
```

The split counts appear to be correct. Let's create our model and see what we get:

```
linearModel <- lm(MDEV ~ CRIM + ZN + INDUS + CHAS + NOX + RM + AGE + DIS +
RAD + TAX + PRATIO + B + LSTAT, data=housingTraining)
summary(linearModel)
```

This is the screenshot of the preceding code:

```
Call:
lm(formula = MDEV ~ CRIM + ZN + INDUS + CHAS + NOX + RM + AGE +
    DIS + RAD + TAX + PRATIO + B + LSTAT, data = housingTraining)

Residuals:
     Min       1Q   Median       3Q      Max
-15.5338  -2.8856  -0.5545   1.9261  25.8096

Coefficients:
              Estimate Std. Error t value Pr(>|t|)
(Intercept)  39.157565   6.094664   6.425 4.10e-10 ***
CRIM         -0.108577   0.035530  -3.056  0.00241 **
ZN            0.034926   0.016350   2.136  0.03333 *
INDUS         0.036136   0.073009   0.495  0.62093
CHAS          2.419065   0.948595   2.550  0.01117 *
NOX         -21.072967   4.513103  -4.669 4.25e-06 ***
RM            3.859993   0.485525   7.950 2.32e-14 ***
AGE           0.005305   0.015526   0.342  0.73280
DIS          -1.445059   0.237213  -6.092 2.82e-09 ***
RAD           0.324186   0.079563   4.075 5.65e-05 ***
TAX          -0.012040   0.004581  -2.628  0.00895 **
PRATIO       -1.021703   0.149298  -6.843 3.26e-11 ***
B             0.008212   0.003204   2.563  0.01078 *
LSTAT        -0.548625   0.060801  -9.023  < 2e-16 ***
---
Signif. codes:  0 '***' 0.001 '**' 0.01 '*' 0.05 '.' 0.1 ' ' 1

Residual standard error: 4.852 on 367 degrees of freedom
Multiple R-squared:  0.7431,    Adjusted R-squared:  0.734
F-statistic: 81.66 on 13 and 367 DF,  p-value: < 2.2e-16
```

It's interesting that several of the variables do not have much of an effect. These are AGE, TAX, and B.

We have our model, so now we can make predictions:

```
predicted <- predict(linearModel,newdata=housingTesting)
summary(predicted)
Min. 1st Qu. Median Mean 3rd Qu. Max.
0.1378 17.4939 21.9724 22.0420 25.6669 40.9981
```

I don't think this summary tells us much. A plot of the two against each other (including an abline function) is much more informative:

```
plot(predicted, housingTesting$MDEV)
abline(lm(predicted ~ housingTesting$MDEV))
```

This is the screenshot of the preceding code:

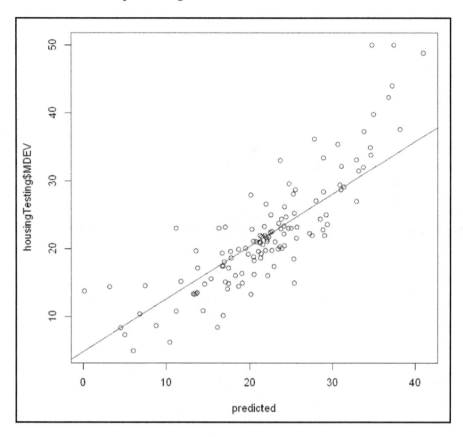

Visually, it looks to be an excellent correlation between the two.

Let's do some math to see how close we got. A sum of squares will give us a good measure. I did not find a built-in method, so I added my own:

```
sumOfSquares <- function(x) {
    return(sum(x^2))
}

#make sure it works
sumOfSquares(1:5)
15
Testing our model:
diff <- predicted - housingTesting$MDEV
sumOfSquares(diff)
2487.85072318584
```

Here, we have a sum of the squares of the differences, which is at about 2,500. This sounds significant for just a few hundred observations.

Summary

In this chapter, we added the ability to use R scripts in our Jupyter notebook. We added an R library that's not included in the standard R installation, and we made a Hello World script in R. We then saw R data access built-in libraries and some of the simpler graphics and statistics that are automatically generated. We used an R script to generate 3D graphics in a couple of different ways. We then performed a standard cluster analysis (which I think is one of the basic uses of R) and used one of the available forecasting tools. We also built a prediction model and tested its accuracy.

In the next chapter, we will learn all about Julia scripting using a Jupyter notebook.

Jupyter Julia Scripting 4

Julia is a language that was specifically designed to be used for high-performance, numerical computing. Most importantly, it differs from the previous scripting languages that have been covered in this book (R and to a certain extent, Python) in that Julia is a full language and not limited to data handling.

In this chapter, we will cover the following topics:

- Adding Julia scripting to your installation
- Basic Julia in Jupyter
- Julia limitations in Jupyter
- Standard Julia capabilities
- Julia visualizations in Jupyter
- Julia Vega plotting
- Julia parallel processing
- Julia control flow
- Julia regular expressions
- Julia unit testing

Adding Julia scripting to your installation

We will install Julia on macOS and Windows. There are very similar steps in both environments due to using Anaconda as the basis for the installation.

Adding Julia scripts to Jupyter

Once Julia is available on your machine, enabling Julia within Jupyter is readily accomplished.

First, we need to install Julia on our Windows machine. Navigate to the Julia download page (http://julialang.org/downloads/), download the correct version, which is Julia 0.6.1 for most environments, and run the installation with the standard default settings.

 You must run the Julia installation as an Administrator on your machine. After downloading the file, open the Downloads folder, right-click on the Julia executable, and select **Run as administrator**.

Once the install is complete, you should verify that everything worked. Select Julia from the programs list and run the Julia program. You should see the command line with Julia displayed, as shown in the following screenshot:

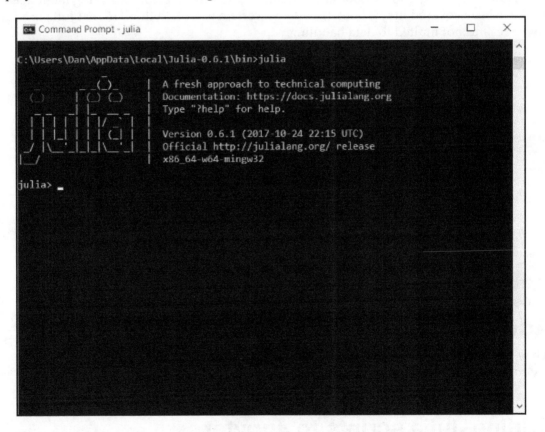

The current version of Julia will not automatically include updates to any packages that may be used. For that, we run the following command to add Julia to Jupyter:

```
Pkg.update()
```

This will result in a number of packages being updated or installed on to your machine. Your display will look something like the following screenshot:

Julia uses font color as feedback. I entered the text `Pkg.update()` in white at the top of the screen; successful execution steps are in blue, and possible problems are shown in red. You must wait for the installation to complete.

This is quite an involved process, where the system looks to see what packages need to be updated and installed, installs each one, verifies that each was successful, and then does it all again until there is nothing left to update.

The last line should have read the following:

```
INFO: Package database updated
```

At this point, you can close the Julia window (by using the `quit()` command).

One last step is to open your Notebook (by using the `jupyter notebook` command), and if you open the **New** menu (in the upper-right corner of the screen), you should see a Julia type available, as shown in the following screenshot:

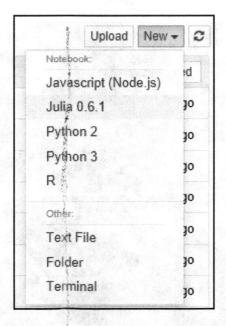

Adding Julia packages to Jupyter

The standard installation for Julia under Jupyter has many packages that are commonly used in Julia programming. However, if you do need to add another package, a small number of steps are required to be followed:

1. Close down your Notebook (including the server).
2. Run the Julia command-line program:

```
Pkg.add("DataFrames")
Pkg.add("RDatasets")
quit();
```

3. Restart your Notebook. The package should be available in your Julia script, for example, `library("name of the package you want to add")`.

I would recommend adding the preceding two packages right away, as they are needed for many scripts.

 The first time you use a package in Julia, you will see a line highlighted in light red that shows Julia is precompiling, such as this: INFO: Precompiling module Dataframes...

You can use the Pkg.add(...) function directly in your script, but that doesn't seem correct. Every time you run your script, the system will attempt to validate whether you have the specified package, install it into your environment if needed, and even tell you whether it is out of date. None of these steps belong to part of your script.

Basic Julia in Jupyter

In this example, we will use the Iris dataset for some standard analysis. So, start a new Julia Notebook and call it Julia Iris. We can enter a small script to see how the steps progress for a Julia script.

This script uses another package for plotting, which is called Gadfly. You will have to go through similar steps as to the ones we went through in the previous section to install the package before operating the script.

Enter the following script into separate cells of your Notebook:

```
using RDatasets
using DataFrames
using Gadfly
set_default_plot_size(5inch, 5inch/golden);
plot(dataset("datasets","iris"), x="SepalWidth", y="SepalLength",
color="Species")
```

RDatasets is a library that contains several of the commonly used R datasets, such as iris. This is a simple script—we define the libraries that we are going to use, set the size of the plot area, and plot out the iris data points (color coded to Species).

So, you will end up with a starting screen that looks like the following screenshot:

I have used **Markdown** cells for the text cells. These serve as documentation of the processing and are not interpreted by the engine.

We should take note of a few aspects of the Julia Notebook view:

- We have the Julia logo (the three colored circles) in the upper-right corner. You will have seen this logo running in other Julia installations (as we saw earlier when we ran the Julia command line previously).
- The circle to the right of the Julia logo is a busy indicator. When your script starts, the title of the table says busy as Julia is starting. When your script is running, the circle is filled in black. When it is not running, it is empty.
- The rest of the menu items are the same as before.

 On my Windows machine, it took quite a while for the Julia Notebook to start for the first time. The **Kernel starting, please wait...** message was displayed for several minutes.

If you run the script (using the **Cell | Run All** menu command), your output should look like what's shown in the following screenshot:

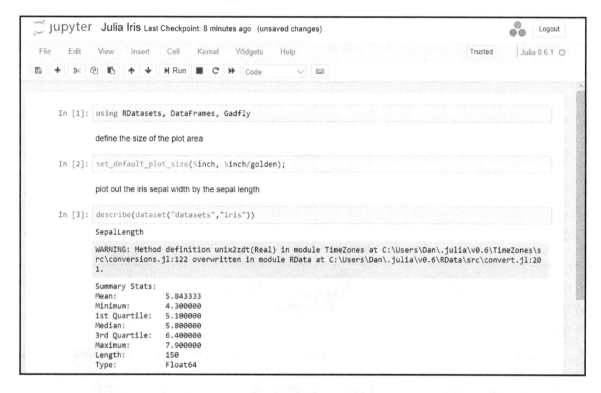

The display continues with other statistics about each of the sets, such as PetalWidth and so on.

Note the WARNING message about an incompatibility between sublibraries. Even with the time it took to install and update packages, there were still unresolved issues.

The more interesting part of this is `plot` of the data points:

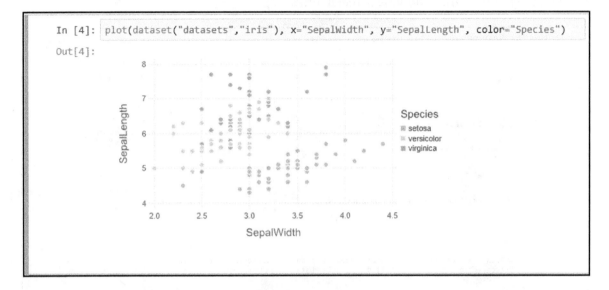

```
In [4]:  plot(dataset("datasets","iris"), x="SepalWidth", y="SepalLength", color="Species")
Out[4]:
```

I noticed that if you hover the mouse over a graphic, you get grid lines displayed and a slide bar to adjust the zoom level (as shown in the upper-right part of the preceding screenshot).

So, just as if you ran the script in the Julia interpreter, you get your output (with the numerical prefix). Jupyter has counted the statements so that we have incremental numbering of the cells. Jupyter has not done anything special to print out variables.

We started the server, created a new Notebook, and saved it as Julia `iris`. If we open the IPYNB file on disk (using a text editor), we can see the following:

```
{
  "cells": [
    ...<similar to previously displayed>
  ],
  "metadata": {
  "kernelspec": {
   "display_name": "Julia 0.6.1",
   "language": "julia",
   "name": "julia-0.6"
  },
  "language_info": {
   "file_extension": ".jl",
   "mimetype": "application/julia",
   "name": "julia",
```

```
    "version": "0.6.1"
  }
},
"nbformat": 4,
"nbformat_minor": 1
}
```

This is a little different than what we saw in the previous chapters with other Notebook language coding. Particularly, `metadata` clearly targets the script cells to be Julia script.

Julia limitations in Jupyter

I have written Julia scripts and accessed different Julia libraries without issue in Jupyter. I have not noticed any limitations on its use or any performance degradation. I imagine some aspects of Julia that are very screen dependent (such as using the **Julia webstack** to build a website) may be hampered by conflicting uses of the same concept.

I have repeatedly seen updates being run when I am attempting to run a Julia script, as shown in the following screenshot. I am not sure why they decided to always update the underlying tool rather than use what is in play and have the user specify whether to update libraries:

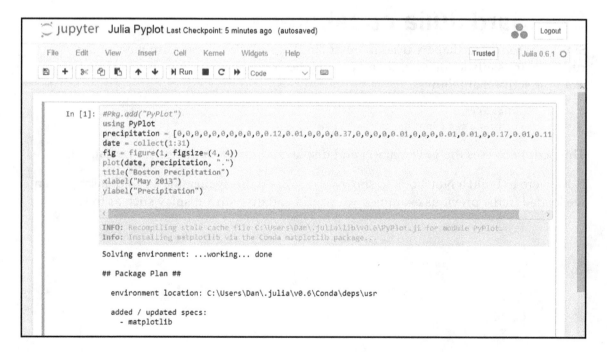

I have also noticed that once a Julia Notebook is opened, even though I have closed the page, it will still display **Running** on the home page. I don't recall seeing this behavior with the other script languages that are available.

Another issue has been trying to use a secured package in my script, for example, `plotly`. It appears to be a clean process to get credentials, but using the prescribed methods for passing your credentials to `plotly` does not work under Windows. I am hesitant to provide examples that do not work in both environments.

Further interactions with Windows are also limited, for example, attempting to access environment variables by calls to standard C libraries that are not normally present on a Windows installation.

I have another issue with Julia itself, regardless of running under Jupyter or not. When using a package, it will complain about features that are used in the package that have been deprecated or improved. As a user of the package, I have no control over this behavior, so it does not help me in my work.

Lastly, running some of these scripts takes several minutes. The scripts used are small. It seems to take a long time for the Julia kernel to start.

Standard Julia capabilities

Similar to functions that are used in other languages, Julia can perform most of the rudimentary statistics on your data by using the `describe` function, as shown in the example script that follows:

```
using RDatasets
describe(dataset("datasets", "iris"))
```

This script accesses the `iris` dataset and displays summary statistics on the dataset.

If we were to build a Notebook to show `describe` in use against the `iris` dataset (which we loaded in the previous example), we would end up with a display such as this:

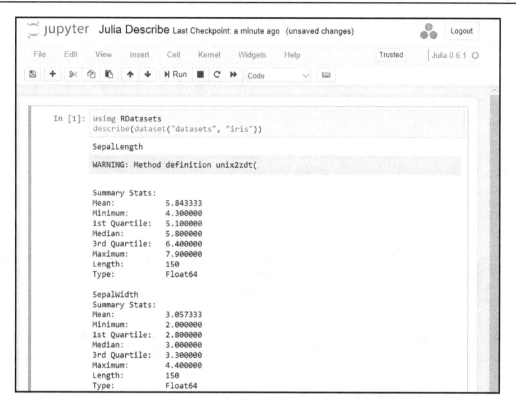

You can see the standard statistics that have been generated for each of the variables in the dataset. I thought it was interesting that the count and percentage of NA values in the dataset are provided. I have found that I usually have to double-check to exclude this data by using other languages. This is a quick, built-in reminder.

 The warning message is complaining about a compatibility issue with one of the datetime libraries used, even though it is not used in this Notebook.

Julia visualizations in Jupyter

The most popular tool for visualizations in Julia is the Gadfly package. We can add the Gadfly package (as described at the beginning of this chapter) by using the add function:

```
Pkg.add("Gadfly")
```

From then on, we can make reference to the `Gadfly` package in any script by using the following:

```
using Gadfly
```

Julia Gadfly scatterplot

We can use the `plot()` function with standard defaults (no type arguments) to generate a scatterplot. For example, with the following simple script:

```
using Gadfly
srand(111)
plot(x=rand(7), y=rand(7))
```

 We use the `srand()` function in all examples that use random results. The `srand()` function sets the random number seed value so that all of the results in this chapter are reproducible.

We generate a nice, clean scatterplot, as shown in the following screenshot:

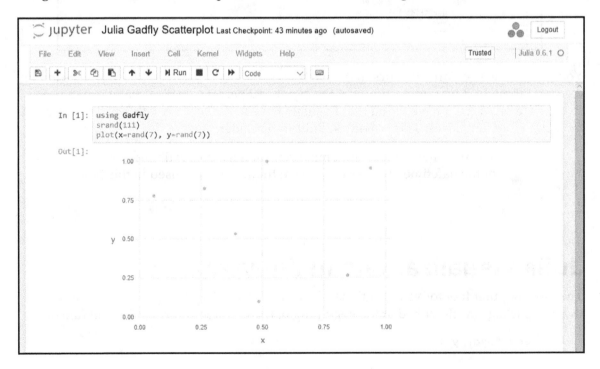

I did notice that if you click on the **?** symbol that appears in the top-right of the graphic if you click on the graphic, a message box is displayed that enables finer control over the graphic to do the following:

- Pan across the image (especially if it expands beyond the window)
- Zoom in, out
- Reset:

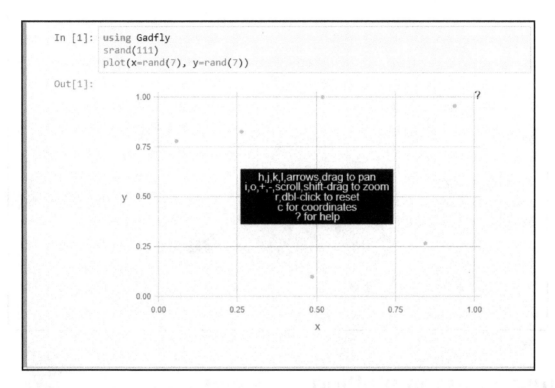

Julia Gadfly histogram

We can produce other graph types as well, for example, histogram, by using the following script:

```
using Gadfly
srand(111)
plot(x=randn(113), Geom.histogram(bincount=10))
```

This script generates 113 random numbers and generates histogram of the results.

We will see something like the following screenshot:

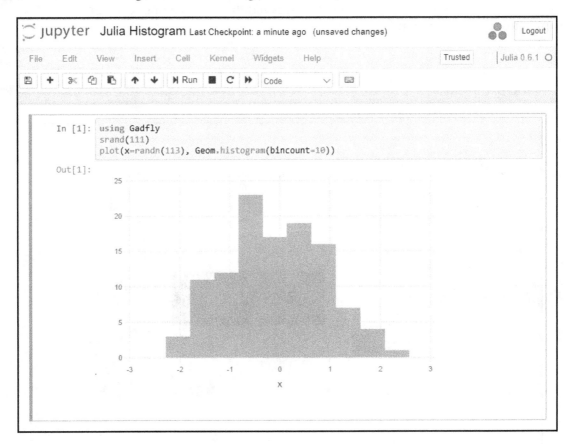

Julia Winston plotting

Another graphics package in Julia is `Winston`. It has similar plotting capabilities to `Gadfly` (I think `Gadfly` is more up-to-date). We can produce a similar plot of random numbers by using the following script:

```
using Winston
# fix the random seed so we have reproducible results
srand(111)
# generate a plot
pl = plot(cumsum(rand(100) .- 0.5), "g", cumsum(rand(100) .- 0.5), "b")
# display the plot
display(pl)
```

Note that you have to specifically display the plot. The `Winston` package assumes that you want to store the graphic as a file, so the `plot` function generates an object for handling.

Moving this into a Notebook, we get the following screenshot:

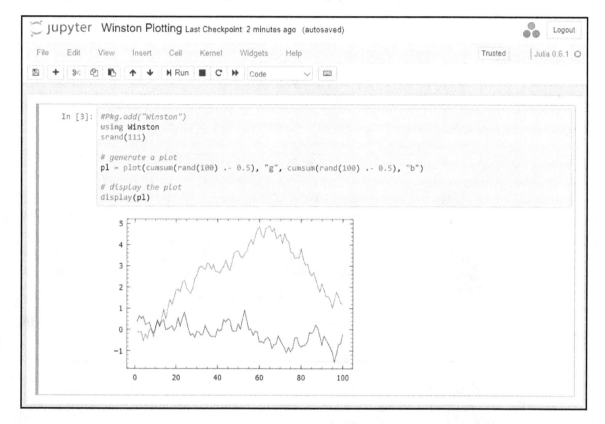

Julia Vega plotting

Another popular graphics package is Vega. The main feature of Vega is the ability to describe your graphic by using language primitives such as JSON. Vega produces most of the standard plots. Here is an example script using Vega for a pie chart:

```
Pkg.add("Vega")
using Vega
stock = ["chairs", "tables", "desks", "rugs", "lamps"];
quantity = [15, 10, 10, 5, 20];
piechart(x = stock, y = quantity)
```

The resultant output in Jupyter may look like the following screenshot:

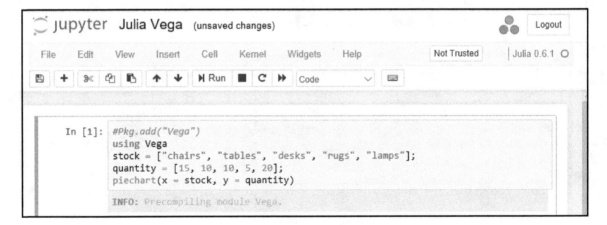

 Note the INFO: Precompiling module Vega. package. Even though the package had been loaded as part of the install or update process, it still needed to adjust the library on first use.

The generated graphic produced under Jupyter is shown in the following screenshot

`Vega` gives you the option on the resultant display to **Save as PNG**. I think this is a useful feature, allowing you to embed the generated graphic(s) in another document:

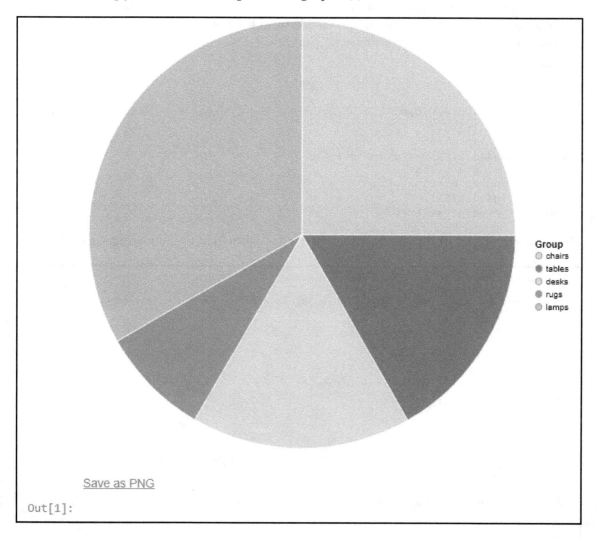

Save as PNG

Out[1]:

Julia PyPlot plotting

Another plotting package available is `PyPlot`. `PyPlot` is one of the standard Python visualization libraries and is directly accessible from Julia. We can take this small script to produce an interesting visualization:

```
#Pkg.add("PyPlot")
using PyPlot
precipitation =
[0,0,0,0,0,0,0,0,0,0,0,0.12,0.01,0,0,0,0.37,0,0,0,0,0.01,0,0,0,0.01,0.01,0,0.
17,0.01,0.11,0.31]
date = collect(1:31)
fig = figure(1, figsize=(4, 4))
plot(date, precipitation, ".")
title("Boston Precipitation")
xlabel("May 2013")
ylabel("Precipitation")
```

The resultant output in Jupyter may look like what's shown in the following screenshot:

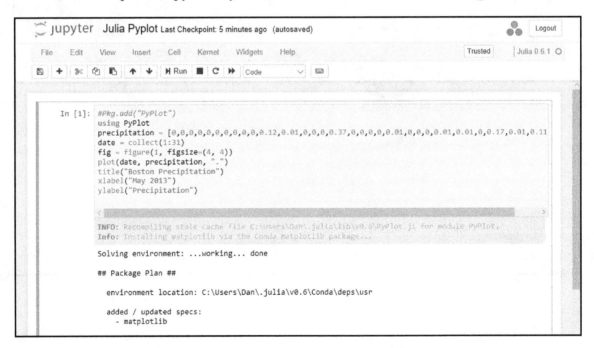

Again, we have Julia updating a package before executing our Notebook until, finally, we get the graphic on `Precipitation`:

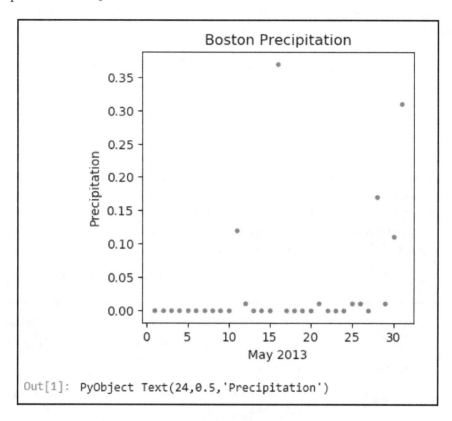

```
Out[1]:  PyObject Text(24,0.5,'Precipitation')
```

It's interesting that Boston has such diverse precipitation—most of the month has none, and then there are a couple of days that have a deluge.

 A reminder: Jupyter will attempt to put most of the output into a small scrolling window. Just clicking in the left-hand side of the display will expand the entire contents of the scroll panel.

Julia parallel processing

An advanced built-in feature of Julia is to use parallel processing in your script. Normally, you can specify the number of processes that you want to use directly in Julia. However, under Jupyter, you would use the `addprocs()` function to add an additional process available for use in your script, for example, look at this small script:

```
addprocs(1)
srand(111)
r = remotecall(rand, 2, 3, 4)
s = @spawnat 2 1 .+ fetch(r)
fetch(s)
```

It makes a call to `rand`, the random number generator, with that code executing on the second parameter to the function call (process 2), and then passes the remaining arguments to the `rand` function there (making rand generate a 3x4 matrix of random numbers). `spawnat` is a macro that evaluates the processes mentioned previously. Then, `fetch` accesses the result of the spawned processes.

We can see the results in the example under Jupyter, as shown in the following screenshot:

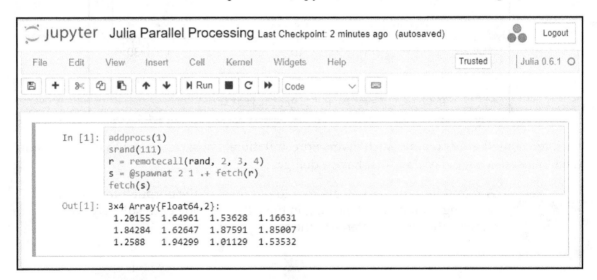

So, this is not a dramatic spawned process type of calculation, but you can easily imagine much more involved processes that are readily available in Jupyter.

Julia control flow

Julia has a complete set of control flows. As an example, we could have a small function `larger` that determines the larger of two numbers:

```
function larger(x, y)
    if (x>y)
        return x
    end
    return y
end
println(larger(7,8))
```

There are several features that you must note:

- The `end` statement for the `if` statement
- `end`, as the closing of the function
- The indentation of the statements within the function
- The indentation of the handling of a true condition within an `if` statement

If we run this under Jupyter, we would see the expected output, as shown in the following screenshot:

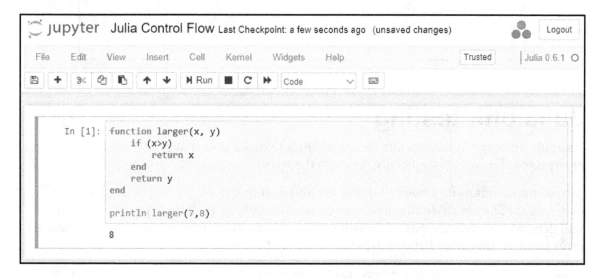

Julia regular expressions

Julia has built-in regular expression handling—as would most modern programming languages. There is no need for using statements, since regular expressions are basic features of strings in Julia.

We could have a small script that validates whether a string is a valid phone number, for example:

```
ismatch(r"^\([0-9]{3}\)[0-9]{3}-[0-9]{4}$", "(781)244-1212")
ismatch(r"^\([0-9]{3}\)[0-9]{3}-[0-9]{4}$", "-781-244-1212")
```

When run under Jupyter, we would see the expected results. The first number is conformant to the format and the second is not:

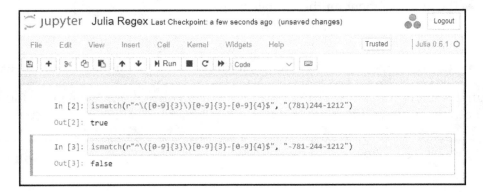

Julia unit testing

As a full language, Julia has unit-testing abilities to make sure that your code is performing as expected. The unit tests usually reside in the tests folder.

Two of the standard functions available for unit testing in Julia are FactCheck and Base.Test. They both do the same thing, but react differently to failed tests. FactCheck will generate an error message that will not stop processing on a failure. If you provide an error handler, that error handler will take control of the test.

Base.Test will throw an exception and stop processing on the first test failure. In that regard, it is probably not useful as a unit-testing function so much as a runtime test that you may put in place to make sure parameters are within reason or otherwise. Just stop processing before something bad happens.

Both packages are built in to the standard Julia distributions.

As an example, we can create a unit tests Notebook that does the same tests and see the resulting, different responses for errors (meaning, test failures).

For FactCheck, we will use the following script:

```
using FactCheck
f(x) = x^3
facts("cubes") do
    @fact f(2) --> 8
    @fact f(2) --> 7
End
```

We are using the FactCheck package. The simple function we are testing is cubing a number, but it could be anything. We wrap our tests in a facts() do...End block. Each of the tests is run within the block which is separate from any other block—so as to group our unit tests together—and is prefixed with @fact. Also, note that we are testing whether the function result following --> is the right-hand argument.

When we run this in Jupyter, we see the expected results, as shown in the following screenshot:

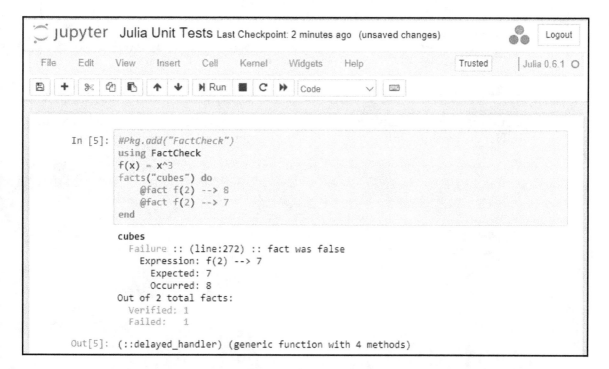

You can see the failed test, why it failed, what line it was on, and so on, as well as the summary for the `facts` block that was executed, that is, the number of tests that passed (`Verified`) and the number of tests that `Failed`. Note that the script continued to run on to the next line.

For `Base.Test`, we have a similar script:

```
using Base.Test f(x) = x^3 @test f(2) == 8 @test f(2) == 7
```

We are using the `Base.Test` package. The function definition we are using is, again, cubing. Then, each test is individually—not as part of a `test` block—prefixed with `@test`. Running this script in Jupyter, we see similar results as the ones that are shown in the following screenshot:

```
In [6]:  using Base.Test
         f(x) = x^3
         @test f(2) == 8
         @test f(2) == 7

         Test Failed
           Expression: f(2) == 7
           Evaluated: 8 == 7

         There was an error during testing

         Stacktrace:
          [1] record(::Base.Test.FallbackTestSet, ::Base.Test.Fail) at .\test.jl:533
          [2] do_test(::Base.Test.Returned, ::Expr) at .\test.jl:352
```

The failed test information is displayed. However, in this case, the script stopped executing at this point. Hence, I would only consider this for runtime checks to validate input formats.

Summary

In this chapter, we added the ability to use Julia scripts in our Jupyter Notebook. We added a Julia library that's not included in the standard Julia installation. We saw basic features of Julia in use, and also outlined some of the limitations that are encountered using Julia in Jupyter. We displayed graphics using some of the graphics packages available, including `Gadfly`, `Winston`, `Vega`, and `PyPlot`. Finally, we saw parallel processing in action, a small control flow example, and how to add unit testing to your Julia script.

In the next chapter, we will learn all about using JavaScript in a Jupyter Notebook.

Jupyter Java Coding 5

Java is a high-level programming language that was originally developed by Sun Microsystems, and is currently owned by Oracle. Java is a cross-platform, compiled language that can be executed on a variety of platforms. Java is cross-platform since it generates **p-code**, which is interpreted by a resident-specific version of Java, the **Java Virtual Machine (JVM)**.

Java is distributed by using a **Java Runtime Executable (JRE)** for those that only need to execute programs that are written. Otherwise, there's a **Java Development Kit (JDK)** for those developing Java applications.

In this chapter, we will cover the following topics:

- Adding the Java kernel to Jupyter
- Java Hello World Jupyter Notebook
- Basic Java in Jupyter

 Major caveat is this does not work on a Windows environment. A Java Notebook will not start on Windows.

Adding the Java kernel to your installation

In this section, we will add the Java kernel to your installation. The steps are very similar, regardless of whether you're installing in a Windows or a macOS environment.

The Java kernel, IJava, was developed and maintained by Spence Park at `https://github.com/SpencerPark/IJava`. There are a couple of requirements for using the Java kernel, which will be covered in the following sections.

Installing Java 9 or later

You can check the version of Java you have installed by using the following command at a command-line prompt:

```
java --version
```

We need version 9 or later.

Also, the installed version must be a JDK. The JRE will not suffice. You can download the latest Java version at www.oracle.com/technetwork/java. At the time of writing this book, version 10 was generally available so I installed that version, as you can see from the following screenshot:

```
Command Prompt
Microsoft Windows [Version 10.0.17134.48]
(c) 2018 Microsoft Corporation. All rights reserved.

C:\Users\Dan>java --version
java 10.0.1 2018-04-17
Java(TM) SE Runtime Environment 18.3 (build 10.0.1+10)
Java HotSpot(TM) 64-Bit Server VM 18.3 (build 10.0.1+10, mixed mode)

C:\Users\Dan>
```

A Jupyter environment is required

This may sound redundant, but this is broad, allowing IJava to run in Jupyter, **JupyterLab**, or **nteract** all Jupyter environments, depending on your needs.

Configuring IJava

Once you have installed Java, you need to configure IJava.

Downloading the IJava project from GitHub

We can download the IJava extension from GitHub by using the following command:

```
> git clone https://github.com/SpencerPark/IJava.git --depth 1
```

```
C:\Users\Dan>git clone https://github.com/SpencerPark/IJava.git --depth 1
Cloning into 'IJava'...
remote: Counting objects: 44, done.
remote: Compressing objects: 100% (31/31), done.
remote: Total 44 (delta 6), reused 31 (delta 4), pack-reused 0
Unpacking objects: 100% (44/44), done.
```

The `git clone` command downloads the project files into the `IJava` directory where you are located (in my case, this is my default user directory):

```
> cd IJava/
```

This command just relocates into the `IJava` directory that was downloaded.

Building and installing the kernel

Following are the commands for the particular operating system:

- ***nix**: chmod u+x gradlew && ./gradlew installKernel
- **Windows**: gradlew installKernel

`gradlew` is a Windows version of Gradle, a popular scripting system. Gradle is adept at installing software. `gradlew` was installed as part of the `git clone` command that you ran earlier:

```
FAILURE: Build failed with an exception.

* What went wrong:
Execution failed for task ':compileJava'.
> Could not target platform: 'Java SE 9' using tool chain: 'JDK 8 (1.8)'.

* Try:
Run with --stacktrace option to get the stack trace. Run with --info or --debug option to get more log output.

* Get more help at https://help.gradle.org

BUILD FAILED in 52s
1 actionable task: 1 executed
```

So, as you can see from the install output, IJava really expects Java 9 to be installed. Java 9 is no longer a supported version from Oracle. We need to configure the tool to use Java 10. In my case, I had previously installed Java and had set the environment variable, JAVA_HOME, to the older version. Changing the environment variable to point to the Java 10 install worked:

```
Select Command Prompt
Microsoft Windows [Version 10.0.17134.48]
(c) 2018 Microsoft Corporation. All rights reserved.

C:\Users\Dan>cd IJava

C:\Users\Dan\IJava>gradlew installKernel
Starting a Gradle Daemon, 1 incompatible Daemon could not be reused, use --status for details

BUILD SUCCESSFUL in 12s
4 actionable tasks: 4 executed
```

Now, when we look at the list of kernels, we can see that Java is available by using the following command:

```
>jupyter kernelspec list
```

After the execution of the preceding command we will see the following results:

```
C:\Users\Dan\IJava>jupyter kernelspec list
Available kernels:
  java          C:\Users\Dan\.ipython\kernels\java
  javascript    C:\Users\Dan\AppData\Roaming\jupyter\kernels\javascript
  julia-0.6     C:\Users\Dan\AppData\Roaming\jupyter\kernels\julia-0.6
  python2       C:\Users\Dan\AppData\Roaming\jupyter\kernels\python2
  ir            C:\Users\Dan\Anaconda3\share\jupyter\kernels\ir
  python3       C:\Users\Dan\Anaconda3\share\jupyter\kernels\python3
```

Available options

As with other Java installations, we can set a number of Java-specific environment variables as desired:

Setting	Default value	Description
IJAVA_VM_OPTS	""	A space delimited list of command-line options that would be passed to the java command if running code. For example, we would use -Xmx128m to set a limit on the heap size or -ea to enable assert statements.
IJAVA_COMPILER_OPTS	""	A space delimited list of command-line options that would be passed to the javac command when compiling a project. For example, -parameters to enable retaining parameter names for reflection.
IJAVA_TIMEOUT	1	A duration in milliseconds, specifying a timeout on long-running code. If less than zero, the timeout is disabled.
IJAVA_CLASSPATH	""	-, a file path separator delimited list of classpath entries that should be available to the user code.
IJAVA_STARTUP_SCRIPTS_PATH	""	A file path seperator delimited list of .jshell scripts to run on startup. This includes ijava-jshell-init.jshell and ijava-magics-init.jshell.
IJAVA_STARTUP_SCRIPT	""	A block of Java code to run when the kernel starts up. This may be something like import my.utils; to set up some default imports or even void sleep(long time) { try {Thread.sleep(time)} catch (InterruptedException e) {}} to declare a default utility method to use in the Notebook.

As you can tell from the preceding descriptions, none of these are required to get a working Java application running. They are normally for special-case handling.

Jupyter Java console

You can run Jupyter in `console` mode, meaning that command lines can be entered directly rather than in a new Notebook in a browser. The command is as follows:

```
jupyter console --kernel=java
```

This means that you can start Jupyter in a console window using the Java kernel. We will see a window like the following one, where we can enter some Java code:

```
)an:        :~ loomeyD$ jupyter console --kernel=java
Jupyter console 5.2.0

Java 10.0.1+10 :: IJava kernel 1.0.8-SNAPSHOT
'rotocol v5.0 implementation by jupyter-jvm-basekernel 2.0.0-SNAPSHOT


[n [1]: String hello = "Hello, Dan"

[n [2]: hello
)ut[2]: "Hello, Dan"
```

Odd interface lines of the command-line interface screen react as if they are part of a Notebook:

```
String hello = "Hello, Dan"
hello
```

But this is not normal Java. There are no semicolons at the end of lines. Semicolons are optional for single-line Java statements.

Also, the single line `hello` is just the reference to the `hello` variable. I am not sure what is causing this to echo into the output.

We can extract this snippet into a Java Notebook with similar results:

Jupyter Java output

The Java implementation is able to differentiate between stdout and stderr, as can be seen with the following small code snippet:

```
System.out.println("stdout");
System.err.println("stderr");
```

When run in a Notebook, the stderr output is colored red:

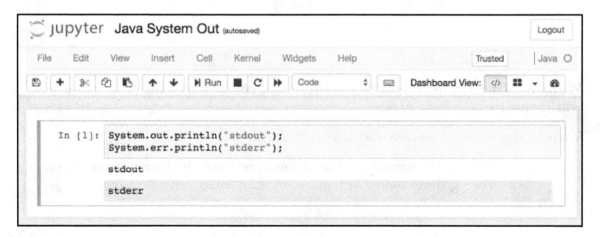

Java Optional

Many a programmer has been bitten by `NullPointerException`. While occurring less in Java than C or C++, it may still occur. Now, Java has the idea of an `Optional` field. An `Optional` field may or may not contain a value. You can test whether there is a value present or not rather than the awkward `null` tests that exist.

We can run through several aspects of `Optional` by using the following code snippet:

```java
import java.util.Optional;
public class MyOptional {
    public static void main() {
        MyOptional program = new MyOptional();
        Integer value1 = null;
        Integer value2 = 123;
        //.ofNullable allows null
        Optional<Integer> a = Optional.ofNullable(value1);
        //.of does not allow null
        Optional<Integer> b = Optional.of(value2);
        System.out.println(program.sum(a,b));
    }
    public Integer sum(Optional<Integer> first, Optional<Integer>
      second) {
        System.out.println("First parameter present " +
          first.isPresent());
        System.out.println("Second parameter present " +
          second.isPresent());
        Integer value1 = first.orElse(1);
        Integer value2 = second.orElse(1);
        return value1 + value2;
    }
}
new MyOptional().main();
```

We have a standard preamble to the class to `import` libraries as needed. In this case, we will just use the `Optional` package.

We will create a Java class with a static `main` function.

 The `main()` function that we defined previously is also nonstandard Java. The signature should be `public static void main(String[] args)`.

First, the `main` function creates an instance of the class (as we will want to reference other parts of the class later on).

We then create two variables, one that is the evil `null` value.

`Optional` has two methods which do the same thing, but behave differently:

- `ofNullable`: Takes an argument, that may be `null`, and creates an `Optional` field
- `of`: Takes an argument, which can not be `null`, and creates an `Optional` field

We now have two `Optional` fields that we pass to the `sum()` function.

The `sum` function uses the `orElse()` function on each `Optional` field, expecting one or both to be `null` and providing a safe passage in those cases.

Then, it is a simple case of mathematics:

```
jupyter  Java Optional  Last Checkpoint: 2 minutes ago  (autosaved)                        Logout

File    Edit    View    Insert    Cell    Kernel    Widgets    Help                Trusted   | Java  O

                      Run        C       Code            Dashboard View: </>

In [8]:  import java.util.Optional;

         public class MyOptional {

             public static void main() {
                 MyOptional program = new MyOptional();

                 Integer value1 = null;
                 Integer value2 = 123;

                 //.ofNullable allows null
                 Optional<Integer> a = Optional.ofNullable(value1);

                 //.of does not allow null
                 Optional<Integer> b = Optional.of(value2);
                 System.out.println(program.sum(a,b));
             }

             public Integer sum(Optional<Integer> first, Optional<Integer> second) {
                 System.out.println("First parameter present " + first.isPresent());
                 System.out.println("Second parameter present " + second.isPresent());
                 Integer value1 = first.orElse(1);
                 Integer value2 = second.orElse(1);
                 return value1 + value2;
             }
         }

In [9]:  new MyOptional .main();

         First parameter present false
         Second parameter present true
         124
```

As you can see in the preceding output, the first parameter is `null`, but due to the `orElse` function, the function continues to process the results.

Java compiler errors

As with any regular Java program, there may be compile-time errors in your coding. Jupyter Java provides similar feedback, with line numbers corresponding to the lines in your Notebook.

For example, when I first entered the snippet for an example that we will look at a little later on in this chapter, there were a couple of errors:

The first error was attempting to call the `sort` function from a `static` method. The second error was attempting to use the wrong function name. Both of these errors are typical types of compile errors you may see when developing Java applications.

Java lambdas

Lambdas provide a clear and concise way to represent one method interface using an expression. Lambdas are usually developed singularly. Lambdas can look very close to earlier Java implementations or completely foreign, as can be seen in the following example. We develop lambdas using more and more terse syntax:

```
In [9]: public class MyLambdas {

            //invoke a function
            private double perform(int a, int b, MyInterface myInterface) {
                return myInterface.someOperation(a, b);
            }

            //define an interface for any operation
            interface MyInterface {
                double someOperation(int a, int b);
            }

            public static void test() {

                MyLambdas tester = new MyLambdas();

                //different ways to express Lambda functions

                //use argument types
                MyInterface powerTyped = (int a, int b) -> Math.pow(a,b);

                //no argument types
                MyInterface powertNotTyped = (a, b) -> Math.pow(a,b);

                //return statement wrapped with curly braces
                MyInterface powerReturnCurlyBraced = (a, b) -> { return Math.pow(a,b); };

                System.out.println("2 ^ 3 = " + tester.perform(2,3,powerTyped));
                System.out.println("2 ^ 3 = " + tester.perform(2,3,powertNotTyped));
                System.out.println("2 ^ 3 = " + tester.perform(2,3,powerReturnCurlyBraced));
            }
        }

        new MyLambdas().test();

        2 ^ 3 = 8.0
        2 ^ 3 = 8.0
        2 ^ 3 = 8.0
```

All three lambdas perform the same step (as can be seen in the preceding output). However, the implementation is progressively more non-Java looking.

Java Collections

Java Collections went through a major rework in the last few releases of Java. You can now use a lambda function to describe your comparison point. If that addressed object has the built-in `compareTo` function (all of the standard Java objects do), then you are done.

In this case, we build a list of strings (country names) and pass that list to the `Collections.sort` routine. The `sort` routine becomes very minor, invoking the built-in `compareTo` functions for `String` in Java:

When we run, we can see the results in a sorted order.

There is likely a way to do this without modifying the passed-in array.

Java streams

Java streams was a significant improvement with Java 8. Now, Java is able to deal with streams of information flow in a functional manner. In this example, we will use stream in several small examples to show the power of the feature.

The code snippet we are using is as follows:

```
public class MyStreams {

    public static void main(String[] args) {
        List<Integer> numbers = new ArrayList<Integer>();
        numbers.add(3);
        numbers.add(-1);
        numbers.add(3);
        numbers.add(17);
        numbers.add(7);
        System.out.println("Numbers greater than 2");
        numbers.stream()
                .filter(number -> number > 2)
                .forEach(number -> System.out.println(number));
        System.out.println("number size = " +
          numbers.stream().count());
        Integer big = numbers.stream().max((n1,n2) ->
          Integer.compare(n1, n2)).get();
        System.out.println("biggest number = " + big);

        Integer small = numbers.stream().min((n1,n2) ->
          Integer.compare(n1, n2)).get();
        System.out.println("smallest number = " + small);
        System.out.println("Sorted");
        numbers.stream()
                .sorted((n1,n2) -> Integer.compare(n1, n2))
                .forEach(number -> System.out.println(number));

        Integer total = numbers.stream()
                .collect(Collectors.summingInt(i -> i))
                .intValue();
        System.out.println("Total " + total);
        String summary = numbers.stream()
                .collect(Collectors.summarizingInt(i -> i))
                .toString();
```

```
            System.out.println("Summary " + summary);
            System.out.println("Squares");
            numbers.stream()
                    .map(i -> i * i)
                    .forEach(i -> System.out.println(i));
            System.out.println("Growth");
            numbers.stream()
                    .flatMap(i -> build(i))
                    .sorted()
                    .forEach(i -> System.out.println(i));
            System.out.println("Distinct growth");
            numbers.stream()
                    .flatMap(i -> build(i))
                    .sorted()
                    .distinct()
                    .forEach(i -> System.out.println(i));
    }
    static Stream<Integer> build(Integer i) {
            List<Integer> t = new ArrayList<Integer>();
            t.add(i);
            t.add(i*i);
            t.add(i*i*i);
            return t.stream();
    }
}
```

The code uses a collection (`stream`) of numbers for several stream operations. Streams have many more functions built-in to them:

1. First, we use `filter` to pick out the elements of interest
2. We use count to find out how many elements are in the stream
3. We use a lambda function to find the smallest element in the stream
4. Next, we use another lambda to sort the stream elements
5. Then, we use collect and use `summingInt` to add up all of the elements
6. A summary is produced of the stream—this is a built-in function of streams
7. Finally, we use `map` and `flatMap` to perform projections (growth) on the stream elements

The coding and output look like the following (I added horizontal lines to break up the output so that it's more readable):

I have cut off the display as the rest is consistent with expectations.

Similarly, in the following output, I have not displayed all of the output:

```
----------------------
Numbers greater than 2
3
3
17
7
----------------------
number size = 5
----------------------
biggest number = 17
----------------------
smallest number = -1
----------------------
Sorted
-1
3
3
7
17
----------------------
Total 29
----------------------
Summary IntSummaryStatistics{count=5, sum=29, min=-1, average=5.800000, max=17}
----------------------
Squares
9
1
9
289
49
----------------------
Growth
-1
-1
1
3
3
7
9
9
17
27
27
```

Java summary statistics

Java can produce summary statistics for a collection. We can retrieve the Iris dataset and put it into a collection before producing the statistics directly.

I have copied the file from `http://archive.ics.uci.edu/ml/machine-learning-databases/iris/iris.data` to make the processing a little smoother.

We read in the Iris data and then call upon collections to produce a summary.

The code for this example is as follows:

```java
import java.io.IOException;
import java.nio.file.FileSystems;
import java.nio.file.Files;
import java.nio.file.Path;
import java.text.DateFormat;
import java.util.ArrayList;
import java.util.List;
import java.util.Map;
import java.util.Optional;
import java.util.regex.Pattern;
import java.util.stream.Collectors;
import java.util.stream.Stream;

public class Iris {
    public Iris(Double sepalLength, Double sepalWidth, Double
     petalLength, Double petalWidth, String irisClass) {
        this.sepalLength = sepalLength;
        this.sepalWidth = sepalWidth;
        this.petalLength = petalLength;
        this.petalWidth = petalWidth;
        this.irisClass = irisClass;
    }
    private Double sepalLength;
    private Double sepalWidth;
    private Double petalLength;
    private Double petalWidth;
    private String irisClass;
    public Double getSepalLength() {
        return this.sepalLength;
    }
    //other getters and setters TBD
}

public class JavaIris {
    public void test() {
```

```
                   //file originally at
    http://archive.ics.uci.edu/ml/machine-learning-databases/iris/iris.data
            Path path = FileSystems
                .getDefault()
                .getPath("/Users/ToomeyD/iris.csv");
            List<Iris> irises = load(path);
            //produce summary statistics for sepal length values
            String sepalLengthSummary = irises.stream()
                .collect(Collectors.summarizingDouble(Iris::getSepalLength))
                .toString();
            System.out.println("\nSepal Length Summary " + sepalLengthSummary);
        }
        public List<Iris> load(Path path) {
            List<Iris> irises = new ArrayList<Iris>();
            try (Stream<String> stream = Files.lines(path)) {
                stream.forEach((line) -> {
                    System.out.println(line);

                    //put each field into array
                    List<String> fields = new ArrayList<String>();
                    Pattern.compile(",")
                        .splitAsStream(line)
                        .forEach((field) -> fields.add(field));

                    //build up the iris values
                    Double sepalLength = new Double(fields.get(0));
                    Double sepalWidth = new Double(fields.get(1));
                    Double petalLength = new Double(fields.get(2));
                    Double petalWidth = new Double(fields.get(3));
                    String irisClass = fields.get(4);
                    Iris iris = new Iris(sepalLength, sepalWidth,
                     petalLength, petalWidth, irisClass);

                    //add to array
                    irises.add(iris);
                });
            } catch (IOException e) {
                e.printStackTrace();
            }

            return irises;
        }
    }

new JavaIris().test();
```

The code is parsing out each row of the `iris` data into an `Iris` object, and is adding that `Iris` object to an array.

The main routine then calls upon the collections to produce a summary.

The coding looks like the following, where we have `Iris` as a separate object:

```
public class Iris {

    public Iris(Double sepalLength, Double sepalWidth, Double petalLength, Double petalWidth, String irisClass) {
        this.sepalLength = sepalLength;
        this.sepalWidth = sepalWidth;
        this.petalLength = petalLength;
        this.petalWidth = petalWidth;
        this.irisClass = irisClass;
    }

    private Double sepalLength;
    private Double sepalWidth;
    private Double petalLength;
    private Double petalWidth;
    private String irisClass;

    public Double getSepalLength() {
        return this.sepalLength;
    }

    //other getters and setters TBD
}
```

Then, the main coding of the routine to read in the flower information and produce statistics is as follows:

```java
public class JavaIris {

    public void test() {

        //file originally at http://archive.ics.uci.edu/ml/machine-learning-databases/iris/iris.data
        Path path = FileSystems
            .getDefault()
            .getPath("/Users/ToomeyD/iris.csv");
        List<Iris> irises = load(path);

        //produce summary statistics for sepal length values
        String sepalLengthSummary = irises.stream()
            .collect(Collectors.summarizingDouble(Iris::getSepalLength))
            .toString();
        System.out.println("\nSepal Length Summary " + sepalLengthSummary);
    }

    public List<Iris> load(Path path) {
        List<Iris> irises = new ArrayList<Iris>();

        try (Stream<String> stream = Files.lines(path)) {
            stream.forEach((line) -> {
                System.out.println(line);

                //put each field into array
                List<String> fields = new ArrayList<String>();
                Pattern.compile(",")
                    .splitAsStream(line)
                    .forEach((field) -> fields.add(field));

                //build up the iris values
                Double sepalLength = new Double(fields.get(0));
                Double sepalWidth = new Double(fields.get(1));
                Double petalLength = new Double(fields.get(2));
                Double petalWidth = new Double(fields.get(3));
                String irisClass = fields.get(4);
                Iris iris = new Iris(sepalLength, sepalWidth, petalLength, petalWidth, irisClass);

                //add to array
                irises.add(iris);
            });
        } catch (IOException e) {
            e.printStackTrace();
        }
```

The tail of the output looks like the following:

```
6.3,2.8,5.1,1.5,Iris-virginica
6.1,2.6,5.6,1.4,Iris-virginica
7.7,3.0,6.1,2.3,Iris-virginica
6.3,3.4,5.6,2.4,Iris-virginica
6.4,3.1,5.5,1.8,Iris-virginica
6.0,3.0,4.8,1.8,Iris-virginica
6.9,3.1,5.4,2.1,Iris-virginica
6.7,3.1,5.6,2.4,Iris-virginica
6.9,3.1,5.1,2.3,Iris-virginica
5.8,2.7,5.1,1.9,Iris-virginica
6.8,3.2,5.9,2.3,Iris-virginica
6.7,3.3,5.7,2.5,Iris-virginica
6.7,3.0,5.2,2.3,Iris-virginica
6.3,2.5,5.0,1.9,Iris-virginica
6.5,3.0,5.2,2.0,Iris-virginica
6.2,3.4,5.4,2.3,Iris-virginica
5.9,3.0,5.1,1.8,Iris-virginica

Sepal Length Summary DoubleSummaryStatistics{count=150, sum=876.500000, min=4.300000, average=5.843333, max=7.900000}
```

This kind of processing is much easier to accomplish in the other engines that are available in Jupyter.

Summary

In this chapter, we saw the steps to install the Java engine into Jupyter. We saw examples of the different output presentations available from Java in Jupyter. Then, we investigated using `Optional` fields. We saw what a compile error looks like in Java in Jupyter. Next, we saw several examples of lambdas. We used collections for several purposes. Lastly, we generated summary statistics for one of the `Iris` dataset points.

In the next chapter, we will look at how to create interactive widgets that can be used in your Notebook.

Summary

In this chapter we...

Jupyter JavaScript Coding

6

JavaScript is a high-level, dynamic, untyped, and interpreted programming language. There are several outgrowth languages that are based on JavaScript. In the case of Jupyter, the underlying JavaScript is really Node.js. Node.js is an event-based framework that uses JavaScript, which can be used to develop large, scalable applications. Note that this is in contrast to the earlier languages covered in this book which are primarily used for data analysis (Python is a general language as well, but has clear aspects that deal with its capabilities of performing data analysis).

In this chapter, we will cover the following topics:

- Adding JavaScript packages to Jupyter
- JavaScript Jupyter Notebook
- Basic JavaScript in Jupyter
- Node.js `d3` package
- Node.js `stats-analysis` package
- Node.js JSON handling
- Node.js `canvas` package
- Node.js `plotly` package
- Node.js asynchronous threads
- Node.js `decision-tree` package

Adding JavaScript scripting to your installation

In this section, we will install JavaScript scripting on macOS and Windows. There are separate steps for getting JavaScript scripting available on your Jupyter installation for each environment. The macOS installation is very clean. The Windows installation still appears to be in flux, and I would expect the following instructions to change over time.

Adding JavaScript scripts to Jupyter on macOS or Windows

I followed the instructions for loading the JavaScript engine for Anaconda from `https://github.com/n-riesco/iJavaScript`. The steps are as follows:

```
conda install nodejs
npm install -g iJavaScript
ijsinstall
```

At this point, starting Jupyter will provide the **JavaScript (Node.js)** engine as a choice, as shown in the following screenshot:

JavaScript Hello World Jupyter Notebook

Once installed, we can attempt the first JavaScript Notebook by clicking on the **New** menu and selecting JavaScript. We will name the Notebook `Hello, World!` and put the following lines in this script:

```
var msg = "Hello, World!"
console.log(msg)
```

This script sets a variable and displays the contents of the variable. After entering the script and running (**Cell** | **Run All**), we will end up with a Notebook screen that looks like the following screenshot:

We should point out some of the highlights of this page:

- We have the now-familiar language logo in the upper-right corner that depicts the type of script in use
- There is output from every line of the Notebook
- More importantly, we can see the true output of the Notebook (following line one) where the string is echoed
- Otherwise, the Notebook looks as familiar as the other types we have seen

If we look at the contents of the Notebook on disk, we can see similar results as well:

```
{
  "cells": [
    <<same format as seen earlier for the cells>>
  ],
  "metadata": {
    "kernelspec": {
      "display_name": "JavaScript (Node.js)",
```

```
      "language": "JavaScript",
      "name": "JavaScript"
    },
    "language_info": {
      "file_extension": ".js",
      "mimetype": "application/JavaScript",
      "name": "JavaScript",
      "version": "8.9.3"
    }
  },
  "nbformat": 4,
  "nbformat_minor": 2
}
```

So, using the same Notebook and the JSON file format, Jupyter provides a different language for use in the Notebook by changing the `metadata` and `language_info` values appropriately.

Adding JavaScript packages to Jupyter

The JavaScript language does not normally install additional packages—it makes reference to other packages via the runtime include directive which is used in your programs. Other packages can be referenced across the network or copied locally into your environment. It is assumed that accessing a library across the network via a CDN is a more efficient and faster mechanism.

However, Node.js adds the required verb to the JavaScript syntax. In this case, your code requires another module to be loaded, which is assumed to be installed in your current environment. To install another module, use `npm`, for example, in the following command:

```
npm install name-of-module
```

This will install the module that's been referenced (including any embedded packages that are required) on your machine so that a required statement will work correctly.

Basic JavaScript in Jupyter

JavaScript, and even Node.js, are not usually noted for data handling, but for application (website) development. This differentiates JavaScript coding in Jupyter from the languages that we covered earlier. However, the examples in this chapter will highlight using JavaScript for application development with data access and analysis features.

JavaScript limitations in Jupyter

JavaScript was originally used to specifically address the need for scripting inside of an HTML page, usually on the client-side (in a browser). As such, it was built to be able to manipulate HTML elements on the page. Several packages have been developed to further this feature, even for creating a web server, especially using extensions such as Node.js.

The use of any of the HTML manipulation and generation features inside of Jupyter runs into a roadblock, since Jupyter expects to control presentation to the user.

Node.js d3 package

The d3 package has data access functionality. In this case, we will read from a tab-separated file and compute an average. Note the use of the underscore variable name for lodash. Variable names starting with an underscore are assumed to be private. However, in this case, it is just a play on the name of the package we are using, which is lodash, or underscore. lodash is also a widely used a utility package.

For this script to execute, I had to do the following:

- Install d3
- Install lodash
- Install isomorphic-fetch (npm install --save isomorphic-fetch es6-promise)
- Import isomorphic-fetch

The script we will use is as follows:

```
var fs = require("fs");
var d3 = require("d3");
var _ = require("lodash");
var _ = require("isomorphic-fetch");

//read and parse the animals file
console.log("Animal\tWeight");
d3.csv("http://www.dantoomeysoftware.com/data/animals.csv", function(data)
{
    console.log(data.name + '\t' + data.avg_weight);
});
```

This assumes that we have previously loaded the fs and d3 packages using npm, as described in the previous script.

For this example, I created a `data` directory on my website, since the URL that we entered is expected to be an absolute URL, and created a CSV file (`animal.csv`) in that directory:

```
Name,avg_weight
Lion,400
Tiger,400
Human,150
Elephant,2000
```

If we load this script into a Notebook and run it, we get the following output, as expected:

It's important to note that the d3 function (many of them, actually) operate asynchronously. In our case, we are just printing every row of the file. You can imagine more elaborate functionality.

Node.js stats-analysis package

The `stats-analysis` package has many of the common statistics that you may want to perform on your data. You will have to install this package using `npm`, as explained previously.

If we had a small set of people's temperatures to work with, we could get some of the statistics on the data readily by using this script:

```
const stats = require("stats-analysis");

var arr = [98, 98.6, 98.4, 98.8, 200, 120, 98.5];

//standard deviation
var my_stddev = stats.stdev(arr).toFixed(2);

//mean
var my_mean = stats.mean(arr).toFixed(2);

//median
var my_median = stats.median(arr);

//median absolute deviation
var my_mad = stats.MAD(arr);

// Get the index locations of the outliers in the data set
var my_outliers = stats.indexOfOutliers(arr);

// Remove the outliers
var my_without_outliers = stats.filterOutliers(arr);

//display our stats
console.log("Raw data is ", arr);
console.log("Standard Deviation is ", my_stddev);
console.log("Mean is ", my_mean);
console.log("Median is ", my_median);
console.log("Median Abs Deviation is " + my_mad);
console.log("The outliers of the data set are ", my_outliers);
console.log("The data set without outliers is ", my_without_outliers);
```

When this script is entered in a Notebook, we get something similar to what's shown in the following screenshot:

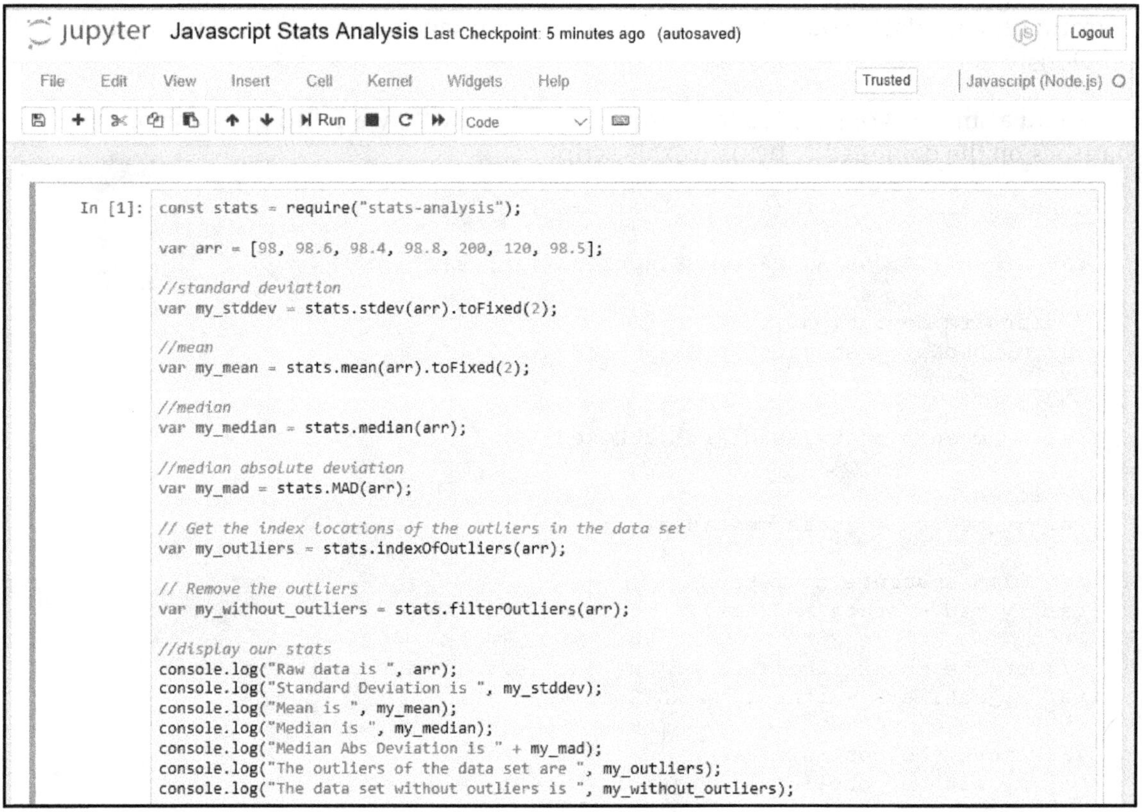

When run, we get the results that are shown in the following screenshot:

```
Raw data is  [ 98, 98.6, 98.4, 98.8, 200, 120, 98.5 ]
Standard Deviation is  35.07
Mean is  116.04
Median is  98.6
Median Abs Deviation is 0.20000000000000284
The outliers of the data set are  [ 4, 5, 6 ]
The data set without outliers is  [ 98, 98.6, 98.4, 98.8 ]
```

Interestingly, 98.5 is considered an outlier. I assume that there is an optional parameter to the command that would change the limits used. Otherwise, the results are as expected.

The outliers are coming from dealing with the raw data as pure mathematical items. So, technically, from the data provided, we have the outliers identified. However, we would likely use a different method to determine outliers, knowing the domain average human temperatures.

Node.js JSON handling

In this example, we will load a JSON dataset and perform some standard manipulations on the data. I am referencing the list of FORD Models from `http://www.carqueryapi.com/api/0.3/?callback=?&cmd=getModels&make=ford`. I can't reference this directly, as it is not a flat file, but an API call. Therefore, I downloaded the data into a local file called `fords.json`. Also, the output from the API call wraps the JSON like so: `?(json);`. This would have to be removed before parsing.

The scripting we will use is as follows. In the script, `JSON` is a built-in package of Node.js, so we can reference this package directly. The `JSON` package provides many of the standard tools that you need to handle your JSON files and objects.

Of interest here is the JSON file reader, which constructs a standard JavaScript array of objects. Attributes of each object can be referenced by `name`, for example, `model.model_name`:

```
//load the JSON dataset
//http://www.carqueryapi.com/api/0.3/?callback=?&cmd=getModels&make=ford
var fords = require('/Users/dtoomey/fords.json');

//display how many Ford models are in our data set
console.log("There are " + fords.Models.length + " Ford models in the data
set");

//loop over the set
var index = 1
for(var i=0; i<fords.Models.length; i++) {
    //get this model
    var model = fords.Models[i];
    //pull it's name
    var name = model.model_name;
    //if the model name does not have numerics in it
    if(! name.match(/[0-9]/i)) {
        //display the model name
        console.log("Model " + index + " is a " + name);
        index++;
    }
```

```
    //only display the first 5
    if (index>5) break;
}
```

If we pull this script into a new Notebook entry, we get the following screenshot:

When the script is executed, we get the expected results, as follows:

```
There are 147 Ford models in the data set
Model 1 is a Aerostar
Model 2 is a Anglia
Model 3 is a Artic
Model 4 is a Aspire
Model 5 is a Bantam
```

Node.js canvas package

The `canvas` package is used for generating graphics in Node.js. We can use the example from the `canvas` package home page (`https://www.npmjs.com/package/canvas`).

First, we need to install `canvas` and its dependencies. There are directions on the home page for the different operating systems, but it is very familiar to the tools we have seen before (we have seen them for macOS):

```
npm install canvas
brew install pkg-config cairo libpng jpeg giflib
```

 This example does not work in Windows. The Windows install required Microsoft Visual C++ to be installed. I tried several iterations to no avail.

With the `canvas` package installed on your machine, we can use a small Node.js script to create a graphic:

```
// create a canvas 200 by 200 pixels
var Canvas = require('canvas')
  , Image = Canvas.Image
  , canvas = new Canvas(200, 200)
  , ctx = canvas.getContext('2d')
  , string = "Jupyter!";

// place our string on the canvas
ctx.font = '30px Impact';
ctx.rotate(.1);
ctx.fillText(string, 50, 100);
var te = ctx.measureText(string);
ctx.strokeStyle = 'rgba(0,0,0,0.5)';
ctx.beginPath();
ctx.lineTo(50, 102);
ctx.lineTo(50 + te.width, 102);
ctx.stroke();
//create an html img tag, with embedded graphics
console.log('<img src="' + canvas.toDataURL() + '" />');
```

This script is creating `canvas`, writing `string` as `Jupyter!` across `canvas`, and then generating an HTML `img` tag with the graphic.

After we run the script in a Notebook, we get the img tag as the output:

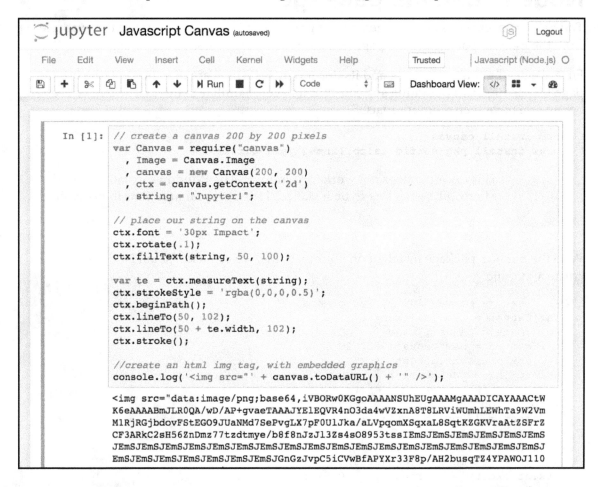

We can take the img tag and save it to an HTML page so it looks like this:

```
<html>
 <body>
 <img src="data:image/png;base64,iVBORw0KGgo<the rest of the tag>CC" />
 </body>
 </head>
```

We can then open the HTML file with a browser to display our graphic:

Node.js plotly package

`plotly` is a package that works differently to most. To use this software, you must register with a `username` so that you are provided with an `api_key` (at https://plot.ly/). You then place the `username` and `api_key` in your script. At that point, you can use all of the `plotly` package features.

First, like all of the other packages, we need to install it:

```
npm install plotly
```

Once installed, we can reference the `plotly` package as needed. Using a simple script, we can generate a `histogram` with `plotly`:

```
//set random seed
var seedrandom = require('seedrandom');
var rng = seedrandom('Jupyter');
//setup plotly
var plotly = require('plotly')(username="<username>", api_key="<key>")
var x = [];
for (var i = 0; i < 500; i ++) {
    x[i] = Math.random();
}
require('plotly')(username, api_key);
var data = [
```

```
    {
      x: x,
      type: "histogram"
    }
  ];
  var graphOptions = {filename: "basic-histogram", fileopt: "overwrite"};
  plotly.plot(data, graphOptions, function (err, msg) {
      console.log(msg);
  });
```

Once loaded and run in Jupyter as a Notebook, we get the following screen:

Instead of creating a local file, or just displaying the graphic on the screen, any graphic that is created is stored on Plotly. The output of the `plot` command is a set of return values from your graphic's creation. Most important is the URL where you can access the graphic.

Ideally, what should happen is that I should be able to access my graphic (histogram) using the URL provided, which is `https://plot.ly/~dantoomey/1`. The returned URL works as expected after inserting a ~ character into the URL. However, when I looked around the Plotly website, I did find my graphics in slightly different paths than expected. All of your graphics are in your home page, which in my case is `https://plot.ly/~dantoomey`. I can now access all of my graphics. The histogram is shown in the following screenshot:

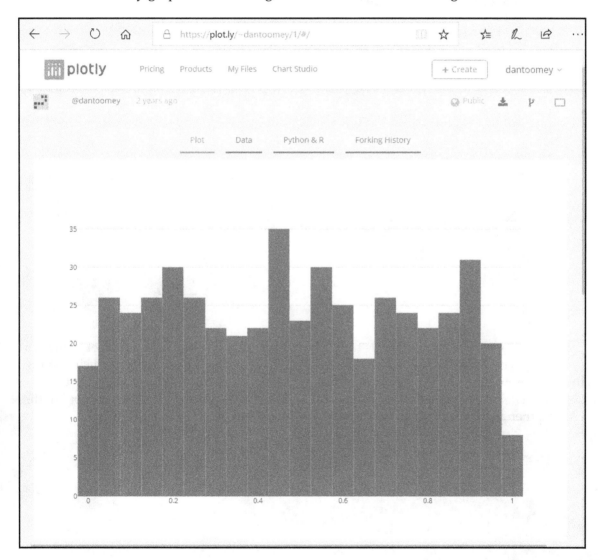

Node.js asynchronous threads

Node.js has built-in mechanisms for creating threads and having them fire asynchronously. Using an example from http://book.mixu.net/node/ch7.html, we have the following:

```
//thread function - invoked for every number in items array
function async(arg, callback) {
  console.log('cube \''+arg+'\', and return 2 seconds later');
  setTimeout(function() { callback(arg * 3); }, 2000);
}

//function called once - after all threads complete
function final() { console.log('Done', results); }

//list of numbers to operate upon
var items = [ 0, 1, 1, 2, 3, 5, 7, 11 ];

//results of each step
var results = [];

//loop the drives the whole process
items.forEach(function(item) {
  async(item, function(result){
    results.push(result);
    if(results.length == items.length) {
      final();
    }
  })
});
```

This script creates an asynchronous function that operates on a number. For every number (item), we call upon the inline function, passing the number to the function which applies the number to the results list. In this case, it just triples the number and waits for two seconds. The main loop (at the bottom of the script) creates a thread for each number in the list and then waits for them all to complete before calling the final() routine.

The Notebook page looks like this:

```
In [1]: //thread function - invoked for every number in items array
        function async(arg, callback) {
          console.log('cube \''+arg+'\', and return 2 seconds later');
          setTimeout(function() { callback(arg * 3); }, 2000);
        }

        //function called once - after all threads complete
        function final() { console.log('Done', results); }

        //list of numbers to operate upon
        var items = [ 0, 1, 1, 2, 3, 5, 7, 11 ];

        //results of each step
        var results = [];

        //loop the drives the whole process
        items.forEach(function(item) {
          async(item, function(result){
            results.push(result);
            if(results.length == items.length) {
              final();
            }
          })
        });
```

When we run the script, we get something like the following output:

```
cube '0', and return 2 seconds later
cube '1', and return 2 seconds later
cube '1', and return 2 seconds later
cube '2', and return 2 seconds later
cube '3', and return 2 seconds later
cube '5', and return 2 seconds later
cube '7', and return 2 seconds later
cube '11', and return 2 seconds later
Done [ 0, 3, 3, 6, 9, 15, 21, 33 ]
```

It is odd to see the delay for the last line of output (from the `final()` routine) to display, although we specifically stated to add a delay when we coded the `async` function previously.

Also, when I played around with other functions, such as cubing each number, the `results` list came back in a very different order. I would not have thought such a basic mathematics function would have any effect on performance.

Node.js decision-tree package

The `decision-tree` package is an example of a machine learning package. It is available at `https://www.npmjs.com/package/decision-tree`. The package is installed by using the following command:

```
npm install decision-tree
```

We need a dataset to use for training/developing our decision tree. I am using the car MPG dataset from the following web page: `https://alliance.seas.upenn.edu/~cis520/wiki/index.php?n=Lectures.DecisionTrees`. It did not seem to be available directly, so I copied it into Excel and saved it as a local CSV.

The logic for machine learning is very similar:

- Load our dataset
- Split into a training set and a testing set
- Use the training set to develop our model
- Test the mode on the test set

 Typically, you might use two-thirds of your data for training and one-third for testing.

Using the `decision-tree` package and the `car-mpg` dataset, we would have a script similar to the following:

```
//Import the modules
var DecisionTree = require('decision-tree');
var fs = require("fs");
var d3 = require("d3");
var util = require('util');

//read in the car/mpg file
fs.readFile("/Users/dtoomey/car-mpg.csv", "utf8", function(error, data) {
    //parse out the csv into a dataset
    var dataset = d3.tsvParse(data);
    //display on screen - just for debugging
    //console.log(JSON.stringify(dataset));

    var rows = dataset.length;
    console.log("rows = " + rows);
    var training_size = rows * 2 / 3;
```

```
    console.log("training_size = " + training_size);
    var test_size = rows - training_size;
    console.log("test_size = " + test_size);

    //Prepare training dataset
    var training_data = dataset.slice(1, training_size);

    //Prepare test dataset
    var test_data = dataset.slice(training_size, rows);

    //Setup Target Class used for prediction
    var class_name = "mpg";

    //Setup Features to be used by decision tree
    var features = ["cylinders","displacement","horsepower", "weight",
"acceleration", "modelyear", "maker"];

    //Create decision tree and train model
    var dt = new DecisionTree(training_data, class_name, features);
    console.log("Decision Tree is " + util.inspect(dt, {showHidden: false,
depth: null}));

    //Predict class label for an instance
    var predicted_class = dt.predict({
        cylinders: 8,
        displacement: 400,
        horsepower: 200,
        weight: 4000,
        acceleration: 12,
        modelyear: 70,
        maker: "US"
    });
    console.log("Predicted Class is " + util.inspect(predicted_class,
{showHidden: false, depth: null}));

    //Evaluate model on a dataset
    var accuracy = dt.evaluate(test_data);
    console.log("Accuracy is " + accuracy);

    //Export underlying model for visualization or inspection
    var treeModel = dt.toJSON();
    console.log("Decision Tree JSON is " + util.inspect(treeModel,
{showHidden: false, depth: null}));
});
```

There is wide use of `console.log` to display progressive information about the processing that is occurring. I am using the `util()` function further in order to display members of objects in use.

The packages must also be installed using npm.

If we run this in a Notebook, we end up with the results that are shown at the top of the following output:

```
rows = 42
training_size = 28
test_size = 14
Decision Tree is { data:
   [ { 'mpg,cylinders,displacement,horsepower,weight,acceleration,modelyear,maker': 'Bad,8,400,170,47
46,12,71,America' },
      { 'mpg,cylinders,displacement,horsepower,weight,acceleration,modelyear,maker': 'Bad,8,400,175,43
85,12,72,America' },
```

Here, the system is just logging the entries that it finds in the file and presenting decision points on the different variables we have assigned. For example, mpg is Bad when cylinders is 8, displacement is 400, and so on.

We arrive at a model for determining whether mpg for a vehicle is acceptable, based on the vehicle's characteristics. In this case, we have a bad predictor, as noted in the results.

Summary

In this chapter, we learned how to add JavaScript to our Jupyter Notebook. We saw some of the limitations of using JavaScript in Jupyter. We had a look at examples of several packages that are typical of Node.js coding, including d3 for graphics, stats-analysis for statistics, built-in JSON handling, canvas for creating graphics files, and plotly, which is used for generating graphics with a third-party tool. We also saw how multi-threaded applications can be developed using Node.js under Jupyter. Lastly, we saw machine learning for developing a decision tree.

In the next chapter, we will see how to create interactive widgets that can be used in your Notebook.

7
Jupyter Scala

Scala has become very popular. It is built on top of Java (so has full interoperability, including resorting to inline Java in your Scala code). However, the syntax is much cleaner and more intuitive, reworking some of the quirks in Java.

In this chapter, we will cover the following topics:

- Installing Scala for Jupyter
- Using Scala features

Installing the Scala kernel

The steps for macOS are as follows (taken from `https://developer.ibm.com/hadoop/2016/05/04/install-jupyter-notebook-spark`):

 I could not get the steps for using the Scala kernel to work on a Windows 10 machine.

1. Install `git` using the following command:

   yum install git

2. Copy the `scala` package locally:

   git clone https://github.com/alexarchambault/jupyter-scala.git

3. Install the `sbt` build tool by running:

   sudo yum install sbt

4. Move the `jupyter-scala` directory to the `scala` package:

   cd jupyter-scala

5. Build the package:

   sbt cli/packArchive

6. To launch Scala shell, use the following command:

   ./jupyter-scala

7. Check the kernels installed by running this command (you should now see Scala in the list):

   jupyter kernelspec list

8. Launch the Jupyter Notebook:

   jupyter notebook

9. You can now choose to use a Scala 2.11 shell.

At this point, if you start Jupyter, you will see **Scala** listed:

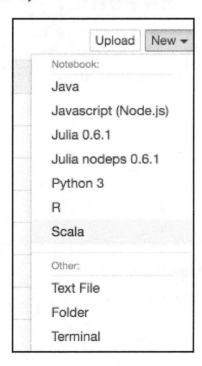

If we create a Scala Notebook, we end up with the familiar layout with an icon displaying that we are running Scala and the engine type string identified as Scala. The kernel name is also specified in the URL to Jupyter:

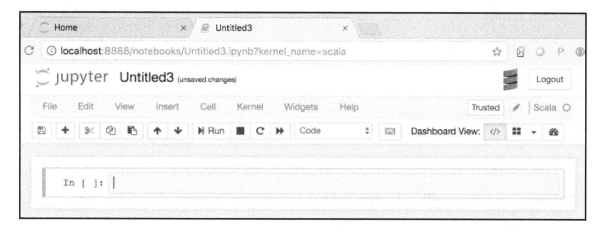

So, after naming our Notebook as Scala Notebook and then saving it, we get the familiar display of Notebooks on the home page, where the new Notebook is called `Scala Notebook.ipynb`.

If we look in the `.ipynb` file, we can see markup similar to other Notebook types, with special markings for Scala:

```
{
 "cells": [
  {
   "cell_type": "code",
   "execution_count": null,
   "metadata": {},
   "outputs": [],
   "source": []
  }
 ],
 "metadata": {
  "kernelspec": {
   "display_name": "Scala",
   "language": "scala",
   "name": "scala"
  },
  "language_info": {
   "codemirror_mode": "text/x-scala",
   "file_extension": ".scala",
   "mimetype": "text/x-scala",
```

```
    "name": "scala211",
    "nbconvert_exporter": "script",
    "pygments_lexer": "scala",
    "version": "2.11.11"
  }
 },
 "nbformat": 4,
 "nbformat_minor": 2
}
```

Now, we can enter Scala coding in some of the cells. Following the previous language examples (from earlier chapters), we can enter the following:

```
val name = "Dan"
val age = 37
show(name + " is " + age)
```

Scala has changeable variables (`var`) and fixed variables (`val`). We are not going to be changing the fields, so they are `val`. The last statement, `show`, is a Jupyter extension for use in Scala to display a variable.

If we run this script in Jupyter, we see the following:

In the output area of the cell, we see the expected `Dan is 37`. Interestingly, Scala also displays the current type and value for each variable in the script at that point as well.

Scala data access in Jupyter

There is a copy of the `iris` dataset on the University of California's website (**Irvine**) at `https://archive.ics.uci.edu/ml/machine-learning-databases/iris/iris.data`. We will access this data and perform several statistical operations on the same:

The Scala code is as follows:

```scala
import scala.io.Source;
//copied file locally
https://archive.ics.uci.edu/ml/machine-learning-databases/iris/iris.data
val filename = "iris.data"
//println("SepalLength, SepalWidth, PetalLength, PetalWidth, Class");
val array = scala.collection.mutable.ArrayBuffer.empty[Float]
for (line <- Source.fromFile(filename).getLines) {
    var cols = line.split(",").map(_.trim);
//println(s"${cols(0)}|${cols(1)}|${cols(2)}|${cols(3)} |${cols(4)}");
    val i = cols(0).toFloat
    array += i;
}
val count = array.length;
var min:Double = 9999.0;
var max:Double = 0.0;
var total:Double = 0.0;
for ( x <- array ) {
    if (x < min) { min = x; }
    if (x > max) { max = x; }
    total += x;
}
val mean:Double = total / count;
```

There seems to be an issue with accessing the CSV file over the internet. So, I copied the file locally, into the same directory where the Notebook resides).

One noteworthy aspect regarding this script is the fact that we do not have to wrap the Scala code in an object, as would normally be required, since Jupyter provides the `wrapper` class.

When we run the script, we see these results:

```
Out[1]:   import scala.io.Source;

          //copied file locally https://archive.ics.uci.edu/ml/machine-learning-dat
          abases/iris/iris.data

          filename: String = "iris.data"
          array: collection.mutable.ArrayBuffer[Float] = ArrayBuffer(
            5.1F,
            4.9F,
            4.7F,
            4.6F,
            5.0F,
            5.4F,
            4.6F,
            5.0F,
            4.4F,
            4.9F,
            5.4F,
          ...
          count: Int = 150
          min: Double = 4.300000190734863
          max: Double = 7.900000095367432
          total: Double = 876.4999990463257
          mean: Double = 5.843333326975505
```

This is a different version of the `iris` data, hence, we see different results in the statistics than we saw earlier.

Scala array operations

Scala does not have pandas, but we can emulate some of that logic with our own coding. We will use the `Titanic` dataset used in `Chapter 2`, *Jupyter Python Scripting*, from `http://titanic-gettingStarted/fdownload/train.csv` which we have downloaded to our local space.

We can then use similar coding to that used in `Chapter 2`, *Jupyter Python Scripting,*
for pandas:

```scala
import scala.io.Source;

val filename = "train.csv"
//PassengerId,Survived,Pclass,Name,Sex,Age,SibSp,Parch,Ticket,Fare,Cabin,Embarked
//1,0,3,"Braund, Mr. Owen Harris",male,22,1,0,A/5 21171,7.25,,S

var males = 0
var females = 0
var males_survived = 0
var females_survived = 0
for (line <- Source.fromFile(filename).getLines) {
    var cols = line.split(",").map(_.trim);
    var sex = cols(5);
    if (sex == "male") {
        males = males + 1;
        if (cols(1).toInt == 1) {
            males_survived = males_survived + 1;
        }
    }
    if (sex == "female") {
        females = females + 1;
        if (cols(1).toInt == 1) {
            females_survived = females_survived + 1;
        }
    }
}
val mens_survival_rate = males_survived.toFloat/males.toFloat
val womens_survival_rate = females_survived.toFloat/females.toFloat
```

In the code, we read the file line by line, parse out the columns (it is CSV), and then make
calculations based on the `sex` column of the data. What is interesting is that Scala arrays are
not zero-based.

When we run this script, we see very similar results to before:

```
jupyter  Scala Array Operations (unsaved changes)                              Logout

File    Edit    View    Insert    Cell    Kernel    Widgets    Help        Trusted    | Scala O

[icons]  Run  ■  C  ▶  Code   ▼  [keyboard]  Dashboard View: </>  ▦  ▼  🜨
```

```scala
//gather statistics on the survivors
var males = 0
var females = 0
var males_survived = 0
var females_survived = 0
for (line <- Source.fromFile(filename).getLines) {
    var cols = line.split(",").map(_.trim);
    var sex = cols(5);
    if (sex == "male") {
        males = males + 1;
        if (cols(1).toInt == 1) {
            males_survived = males_survived + 1;
        }
    }
    if (sex == "female") {
        females = females + 1;
        if (cols(1).toInt == 1) {
            females_survived = females_survived + 1;
        }
    }
}

//see who ended up with higher survival rates
val mens_survival_rate = males_survived.toFloat/males.toFloat
val womens_survival_rate = females_survived.toFloat/females.toFloat
```

```
Out[1]:  import scala.io.Source;

         //load the training data from the titanic survivor set

         filename: String = "train.csv"
         males: Int = 577
         females: Int = 314
         males_survived: Int = 109
         females_survived: Int = 233
         mens_survival_rate: Float = 0.18890814F
         womens_survival_rate: Float = 0.7420382F
```

So, we see drastically higher survival rates for women. I think the stories about women and children first are true.

Scala random numbers in Jupyter

In this example, we simulate rolling dice and counting how many times each combination appears. We then present a simple histogram for illustrative purposes.

The script is as follows:

```scala
val r = new scala.util.Random
r.setSeed(113L)
val samples = 1000
var dice = new Array[Int](12)
for( i <- 1 to samples){
    var total = r.nextInt(6) + r.nextInt(6)
    dice(total) = dice(total) + 1
}
val max = dice.reduceLeft(_ max _)
for( i <- 0 to 11) {
    var str = ""
    for( j <- 1 to dice(i)/3) {
        str = str + "X"
    }
    print(i+1, str, "\n")
}
```

We first pull in the Scala Random library. We set the seed (in order to have repeatable results). We are drawing 1000 rolls. For each roll, we increment a counter of how many times the total number of pips on die one and die two appear. We then present an abbreviated histogram of the results.

Scala has a number of shortcut methods for quickly scanning through a list/collection, as seen in the reduceLeft(_ max _) statement. We can also find the minimum value by using min instead of max in the reduceLeft statement.

When we run the script, we get these results:

```
In [1]:  //setup the random number generator
         val r = new scala.util.Random
         r.setSeed(113L)

         //roll dice 1000 times
         val samples = 1000
         var dice = new Array[Int](12)

         //track all in a histogram
         for( i <- 1 to samples){
             var total = r.nextInt(6) + r.nextInt(6)
             dice(total) = dice(total) + 1
         }
         val max = dice.reduceLeft(_ max _)
         for( i <- 0 to 11) {
             var str = ""
             for( j <- 1 to dice(i)/3) {
                 str = str + "X"
             }
             print(i+1, str, "\n")
         }

         (1,XXXXXXX,
         )(2,XXXXXXXXXXXXXXX,
         )(3,XXXXXXXXXXXXXXXXXXXXXXXX,
         )(4,XXXXXXXXXXXXXXXXXXXXXXXXXXXXXXXXXXXXX,
         )(5,XXXXXXXXXXXXXXXXXXXXXXXXXXXXXXXXXXXXXXXXXXXXXXXX,
         )(6,XXXXXXXXXXXXXXXXXXXXXXXXXXXXXXXXXXXXXXXXXXXXXXXXXXXXXX,
         )(7,XXXXXXXXXXXXXXXXXXXXXXXXXXXXXXXXXXXXXXXXXXXXXXX,
         )(8,XXXXXXXXXXXXXXXXXXXXXXXXXXXXXXXXXXXXXXX,
         )(9,XXXXXXXXXXXXXXXXXXXXXXXXXX,
         )(10,XXXXXXXXXXXXXXXXXXXXXXXX,
         )(11,XXXXXXXXXX,
         )(12,,
         )

Out[1]:  r: util.Random = scala.util.Random@741dc386
         samples: Int = 1000
         dice: Array[Int] = Array(23, 48, 76, 112, 148, 164, 135, 114, 78, 71, 31
         , 0)
         max: Int = 164
```

We can see the crude histogram and the follow-on display of the current values of scalar variables in the script. Do note that I divided by three so that the results would fit on one page.

Scala closures

A closure is a function. The function resultant value depends on the value of variable(s) declared outside the function.

We can use this small script by way of illustration:

```scala
var factor = 7
val multiplier = (i:Int) => i * factor
val a = multiplier(11)
val b = multiplier(12)
```

We define a function named `multiplier`. The function expects an `Int` argument. For each argument, we take the argument and multiply it by the external `factor` variable.

We see the following result:

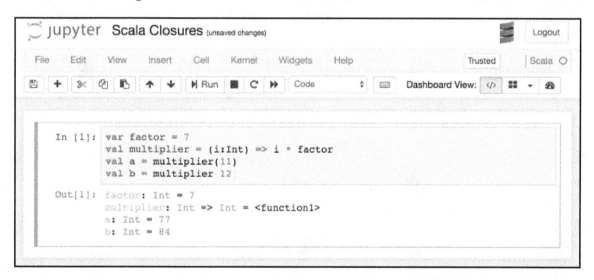

Closures have a nice feel as they do what you expect with little fanfare.

Scala higher-order functions

A higher-order function either takes other functions as arguments or returns a function as its result.

We can use this example script:

```
def squared(x: Int): Int = x * x
def cubed(x: Int): Int = x * x * x

def process(a: Int, processor: Int => Int): Int = {processor(a) }

val fiveSquared = process(5, squared)
val sevenCubed = process(7, cubed)
```

We define two functions: one squares the number passed, and the other cubes the number passed.

Next, we define a higher-order function that takes a number to work on and a processor to apply.

Lastly, we call each function. For example, we call process() with 5 and the squared function. The process() function passes 5 to the squared() function and returns the result:

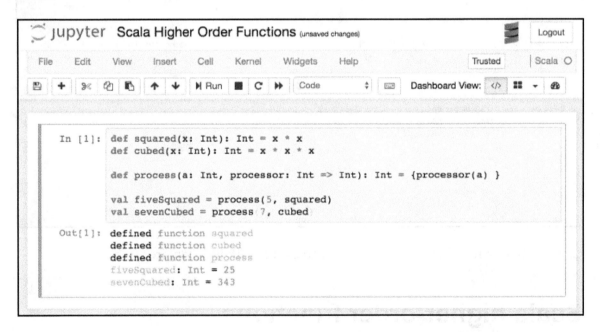

We take advantage of Scala's engine automatically printing out variable values to see the expected result.

These functions are not doing very much. When I ran them, it took a few seconds for the results to display. I think there is a big performance hit using higher-order functions in Scala.

Scala pattern matching

Scala has very useful, built-in pattern matching. Pattern matching can be used to test for exact and/or partial matches of entire values, parts of objects, you name it.

We can use this sample script for reference:

```
def matchTest(x: Any): Any = x match {
  case 7 => "seven"
  case "two" => 2
  case _ => "something"
}
val isItTwo = matchTest("two")
val isItTest = matchTest("test")
val isItSeven = matchTest(7)
```

We define a function called matchTest. The matchTest function takes any kind of argument and can return any type of result. (Not sure if that is real-life programming).

The keyword of interest is match. This means the function will walk down the list of choices until it gets a match on to the x value passed and then returns.

As you can see, we have numbers and strings as input and output.

The last case statement is a wildcard, and _ is catchall, meaning that if the code gets that far, it will match any argument.

We can see the output as follows:

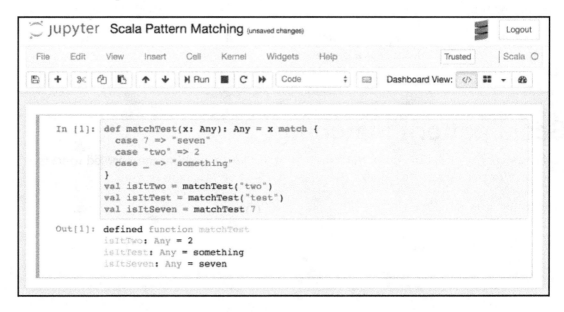

Scala case classes

A `case` class is a simplified type that can be used without calling `new Classname(..)`. For example, we could have this script that defines a `case` class and uses it:

```
case class Car(brand: String, model: String)
val buickLeSabre = Car("Buick", "LeSabre")
```

So, we have a `case` class called `Car`. We make an instance of that class called `buickLeSabre`.

The `case` classes are most useful for pattern matching, since we can easily construct complex objects and examine their content. For example:

```
def carType(car: Car) = car match {
  case Car("Honda", "Accord") => "sedan"
  case Car("GM", "Denali") => "suv"
  case Car("Mercedes", "300") => "luxury"
  case Car("Buick", "LeSabre") => "sedan"
  case _ => "Car: is of unknown type"
}
val typeOfBuick = carType(buickLeSabre)
```

We define a pattern `match` block (as in the previous section of this chapter). In the `match` block, we look at a `Car` object that has `brand` as `GM` and `model` as `Denali`, and so forth. For each of the models of interest, we classify its type. We also have the `catchall` `_` at the end so we can catch unexpected values.

We can exercise the `case` classes in Jupyter, as shown in this screenshot:

We defined and used the `case` class as `Car`. We then did pattern matching using the `Car` class.

Scala immutability

Immutable means you cannot change something. In Scala, all variables are immutable, unless specifically marked otherwise. This is the opposite to languages such as Java, where all variables are mutable unless specifically marked otherwise.

In Java, we can have the following function:

```
public void calculate(integer amount) {
}
```

We can modify the value of `amount` inside the `calculate` function. We can tell Java not to allow changing the value if we use the `final` keyword, as in:

```
public void calculate(final integer amount) {
}
```

Whereas in Scala:

```
def calculate (amount: Int): Int = {
        amount = amount + 1;
        return amount;
}
var balance = 100
val result = calculate(balance)
```

A similar routine leaves the value of the `amount` variable as it was before the routine was called:

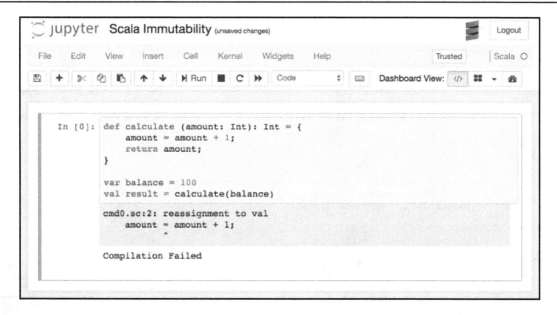

We can see in the display that even though `balance` is a variable (marked as `var`), Scala will not allow you to change its value inside the function. The `amount` parameter to the `calculate` function is assumed to be a `val` and cannot be changed once initialized.

Scala collections

In Scala, collections are automatically `mutable` or `immutable` depending on your usage. All collections in `scala.collections.immutable` are `immutable`. And vice-versa for `scala.collections.immutable`. Scala picks `immutable` collections by default, so your code will then draw automatically from the `mutable` collections, as in:

```
var List mylist;
```

Or, you can prefix your variable with `immutable`:

```
var mylist immutable.List;
```

We can see this in this short example:

```
var mutableList = List(1, 2, 3);
var immutableList = scala.collection.immutable.List(4, 5, 6);
mutableList.updated(1,400);
immutableList.updated(1,700);
```

As we can see in this Notebook:

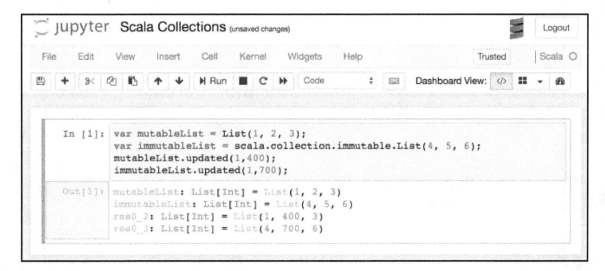

Note that Scala cheated a little here: it created a new `collection` when we updated `immutableList`, as you can see with the variable name `real_3` instead.

Named arguments

Scala allows you to specify parameter assignment by name rather than just ordinal position. For example, we can have this code:

```
def divide(dividend:Int, divisor:Int): Float =
{ dividend.toFloat / divisor.toFloat }
divide(40, 5)
divide(divisor = 40, dividend = 5)
```

If we run this in a Notebook, we can see the results:

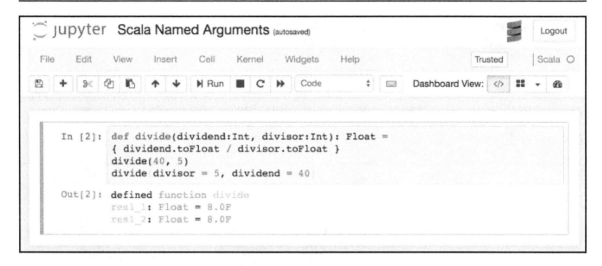

The first call divides the assigned parameters by position. The second call sets the parameters specifically, rather than using standard ordinal position assignment.

Scala traits

The `trait` function in Scala defines a set of features that can be implemented by classes. A `trait` interface is similar to an interface in Java.

The `trait` function can be partially implemented, forcing the user (class) of `trait` to implement the details.

By way of an example, we could have this code:

```
trait Color {
 def isRed(): Boolean
}
class Red extends Color {
 def isRed() = true
}
class Blue extends Color {
 def isRed() = false
}
var red = new Red();
var blue = new Blue();
red.isRed()
blue.isRed()
```

The code creates a trait called Color, with one partially implemented function, isRed. So, every class that uses Color will have to implement isRed().

We then implement two classes, Red and Blue, that extend the Color trait (this is the Scala syntax for using trait). Since the isRed() function is partially implemented, both classes have to provide implementations for the trait function.

We can see how this operates in the following Notebook display:

We see (in the output section at the bottom) that trait and class are created, and that two instances are created, along with the result of calling the trait function for both classes.

Summary

In this chapter, we installed Scala for Jupyter. We used Scala coding to access larger datasets. We saw how Scala can manipulate arrays. We generated random numbers in Scala. There were examples of higher-order functions and pattern matching. We used `case` classes. We saw examples of immutability in Scala. We built collections using Scala packages. Finally, we looked at Scala traits.

In the next chapter, we will be looking at using big data in Jupyter.

8
Jupyter and Big Data

Big data is the topic on everyone's mind. I thought it would be good to see what can be done with big data in Jupyter. One up-and-coming language for dealing with large datasets is Spark. Spark is an open source toolset. We can use Spark coding in Jupyter much like the other languages we have seen.

In this chapter, we will cover the following topics:

- Installing Spark for use in Jupyter
- Using Spark's features

Apache Spark

One of the tools we will be using is Apache Spark. Spark is an open source toolset for cluster computing. While we will not be using a cluster, typical usage for Spark is a larger set of machines or clusters that operate in parallel to analyze a big dataset. Installation instructions are available at `https://www.dataquest.io/blog/pyspark-installation-guide`.

Installing Spark on macOS

Up-to-date instructions for installing Spark are available at `https://medium.freecodecamp.org/installing-scala-and-apache-spark-on-mac-os-837ae57d283f`. The main steps are:

1. Get Homebrew from `http://brew.sh`. If you are doing software development on macOS, you will likely already have Homebrew.

2. Install `xcode-select`: `xcode-select` is used for different languages. For Spark we use Java, Scala and, of course, Spark as follows:

   ```
   xcode-select -install
   ```

   Again, it is likely that you will already have this for other software development tasks.

3. Use Homebrew to install Java:

   ```
   brew cask install java
   ```

4. Use Homebrew to install Scala:

   ```
   brew install scala
   ```

5. Use Homebrew to install Apache Spark:

   ```
   brew install apache-spark
   ```

6. You should test whether this is working using the Spark shell, as in this command:

   ```
   spark-shell
   ```

7. This will result in the familiar logo display:

```
Welcome to
      ____              __
     / __/__  ___ _____/ /__
    _\ \/ _ \/ _ `/ __/  '_/
   /__ / .__/\_,_/_/ /_/\_\   version 2.0.0
      /_/

Using Python version 2.7.12 (default, Jul  2 2016 17:43:17)
SparkSession available as 'spark'.
>>>
```

The site continues on to talk about setting up exports and the like, but I did not need to do this.

At this point, we can bring up a Python 3 Notebook and start using Spark in Jupyter.

You can type `quit()` to exit.

Now, when we run our Notebook while using a Python kernel, we can access Spark.

Windows install

I had run Spark in Python 2 in previous of Jupyter. I could not get the installs to work correctly for Windows.

First Spark script

Our first script reads in a text file and sees how much the line lengths add up to, as shown next. Note that we are reading in the Notebook file we are running; the Notebook is named `Spark File Lengths`, and is stored in the `Spark File Lengths.ipynb` file:

```
import pyspark
if not 'sc' in globals():
    sc = pyspark.SparkContext()
lines = sc.textFile("Spark File Line Lengths.ipynb")
lineLengths = lines.map(lambda s: len(s))
totalLengths = lineLengths.reduce(lambda a, b: a + b)
print(totalLengths)
```

In the `print(totalLengths)` script, we first initialize Spark, but only if we have not done so already. Spark will complain if you try to initialize it more than once, so all Spark scripts should have this `if` statement prefix.

The script reads in a text file (the source of this script), takes every line and computes its length, and then adds all the lengths together.

A `lambda` function is an anonymous (not named) function that takes arguments and returns a value. In the first case, given a `s` string return its length.

A `reduce` function takes each value as an argument, applies the second argument to it, replaces the first value with the result, and then proceeds with the rest of the list. In our case, it walks through the line lengths and adds them all up.

Then, running this in a Notebook, we see the following screenshot. Note that the size of your file may be slightly different.

Also, the first time you begin the Spark engine (using the line `sc = pyspark.SparkContext()`), it may take a while and your script may not complete successfully. If that happens, just try it again:

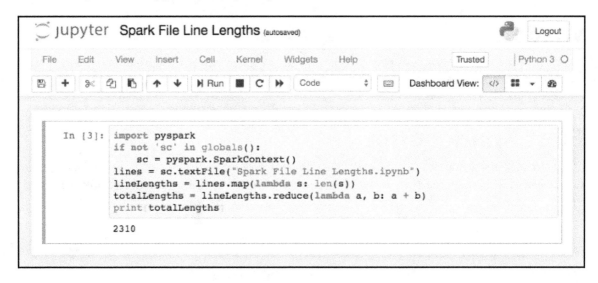

Spark word count

Now that we have seen some of the functionality, let's explore further. We can use a script similar to the following to count the word occurrences in a file:

```
import pyspark
if not 'sc' in globals():
    sc = pyspark.SparkContext()
#load in the file
text_file = sc.textFile("Spark File Words.ipynb")

#split file into distinct words
counts = text_file.flatMap(lambda line: line.split(" ")) \
    .map(lambda word: (word, 1)) \
    .reduceByKey(lambda a, b: a + b)
# print out words found
for x in counts.collect():
    print(x)
```

We have the same preamble to the coding. Then, we load the text file into memory.

Once the file is loaded, we split each line into words and use a `lambda` function to tick off each occurrence of a word. The code is truly creating a new record for each word occurrence, such as at appears one. The idea is that this process could be split over multiple processors, where each processor generates these low-level information bits. We are not concerned with optimizing this process at all.

Once we have all of these records, we reduce/summarize the record set according to the word occurrences mentioned.

The `counts` object is called a **Resilient Distributed Dataset** (**RDD**) in Spark. It is resilient since care is taken to persist the dataset. The RDD is distributed as it can be manipulated by all nodes in the operating cluster. And, of course, it is a dataset consisting of a variety of data items.

The last `for` loop runs `collect()` against the RDD. As mentioned, this RDD could be distributed among many nodes. The `collect()` function pulls all copies of the RDD into one location. Then, we loop through each record.

When we run this in Jupyter, we see something akin to this displayed:

```
In [2]: import pyspark
        if not 'sc' in globals():
            sc = pyspark.SparkContext()

        #load in the file
        text_file = sc.textFile("Spark File Words.ipynb")

        #split file into distinct words
        counts = text_file.flatMap(lambda line: line.split(" ")) \
            .map(lambda word: (word, 1)) \
            .reduceByKey(lambda a, b: a + b)

        # print out words found
        for x in counts.collect():
            print(x)
```

```
('', 204)
('"cells":', 1)
('[', 4)
('"code",', 1)
('1,', 1)
('"outputs":', 1)
('"IndentationError",', 1)
('"evalue":', 1)
('"unexpected', 1)
('(<ipython-input-1-6243b2f81349>,', 1)
('line', 2)
('"output_type":', 1)
('"error",', 1)
('"\\u001b[0;36m', 1)
('\\u001b[0;32m\\"<ipython-input-1-6243b2f81349>\\"\\u001b[0;36m,', 1)
('word:', 2)
('(word,', 2)
('1))\\u001b[0m\\n\\u001b[0m', 1)
('^\\u001b[0m\\n\\u001b[0;31mIndentationError\\u001b[0m\\u001b[0;31m:\\u0
```

The listing is abbreviated as the list of words continues for some time. What is curious is that the word splitting logic in Spark does not appear to work very well; some of the results are not words, such as the first entry an empty string.

Sorted word count

Using the same script with a minor modification, we can make one more call and sort the results. The script now looks as follows:

```python
import pyspark
if not 'sc' in globals():
    sc = pyspark.SparkContext()
#load in the file
text_file = sc.textFile("Spark Sort Words from File.ipynb")

#split file into sorted, distinct words
sorted_counts = text_file.flatMap(lambda line: line.split(" ")) \
    .map(lambda word: (word, 1)) \
    .reduceByKey(lambda a, b: a + b) \
    .sortByKey()
# print out words found (in sorted order)
for x in sorted_counts.collect():
    print(x)
```

Here, we have added another function call to RDD creation, sortByKey(). So, after we have mapped/reduced, and arrived at a list of words and occurrences, we can then easily sort the results.

The resultning output looks like this:

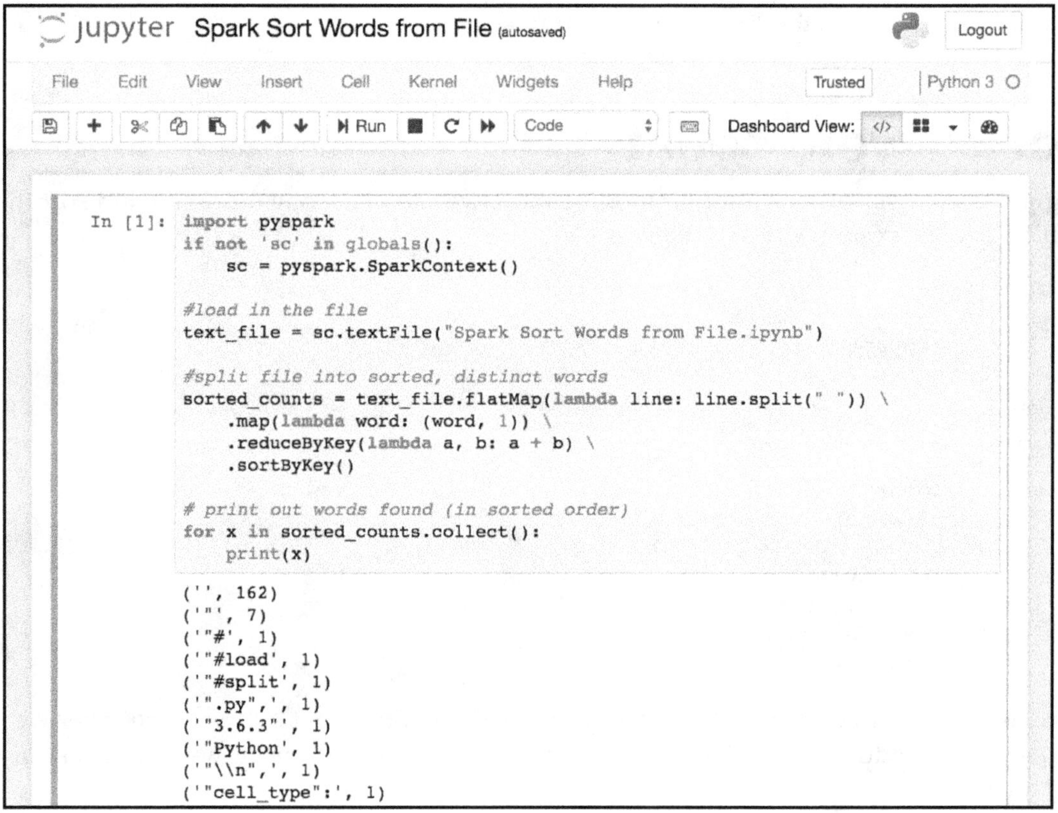

Estimate pi

We can use `map` or `reduce` to estimate pi if we have code like this:

```
import pyspark
import random
if not 'sc' in globals():
    sc = pyspark.SparkContext()
NUM_SAMPLES = 10000
random.seed(113)

def sample(p):
    x, y = random.random(), random.random()
    return 1 if x*x + y*y < 1 else 0
```

```
count = sc.parallelize(range(0, NUM_SAMPLES)) \
     .map(sample) \
     .reduce(lambda a, b: a + b)
print("Pi is roughly %f" % (4.0 * count / NUM_SAMPLES))
```

This code has the same preamble. We are using the Python `random` package. There is a constant for the number of samples to attempt.

We are building an RDD called `count`. We call the `parallelize` function to split this process between the nodes available. The code just maps the result of the `sample` function call. Finally, we reduce the generated map set by adding all the samples.

The `sample` function gets two random numbers and returns a one or a zero depending on where the two numbers end up in size. We are looking for random numbers in a small range and then checking whether they occur within a circle of the same diameter. With a large enough sample, we would end up with `PI (3.141...)`.

If we run this in Jupyter, we see the following:

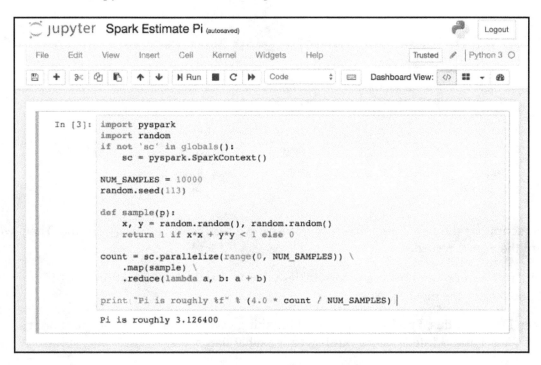

When I ran this with `NUM_SAMPLES = 100000`, I ended up with `PI = 3.126400`.

Log file examination

I downloaded one of the `access_log` files from `monitorware.com`. Like any other web access log, we have one line per entry, like this:

```
64.242.88.10 - - [07/Mar/2004:16:05:49 -0800] "GET
/twiki/bin/edit/Main/Double_bounce_sender?topicparent=Main.Configur
ationVariables HTTP/1.1" 401 12846
```

The first part is the IP address of the caller, followed by a timestamp, the type of HTTP access, the URL referenced, the HTTP type, the resulting HTTP response code, and finally the number of bytes in the response.

We can use Spark to load in and parse out some statistics of the log entries, as in this script:

```
import pyspark
if not 'sc' in globals():
    sc = pyspark.SparkContext()
textFile = sc.textFile("access_log")
print(textFile.count(), "access records")

gets = textFile.filter(lambda line: "GET" in line)
print(gets.count(), "GETs")

posts = textFile.filter(lambda line: "POST" in line)
print(posts.count(), "POSTs")

other = textFile.subtract(gets).subtract(posts)
print(other.count(), "Other")

#curious what Other requests may have been
for x in other.collect():
    print(x)
```

This script has the same preamble as the others. We read in the `access_log` file. Then, we print the `count` record.

Similarly, we find out how many log entries were GET and POST operations. GET is assumed to be the most prevalent.

When I first did this, I really didn't expect anything else, so I removed `gets` and `posts` from the set and printed out the outliers to see what they were.

When we run this in Jupyter, we see the expected output:

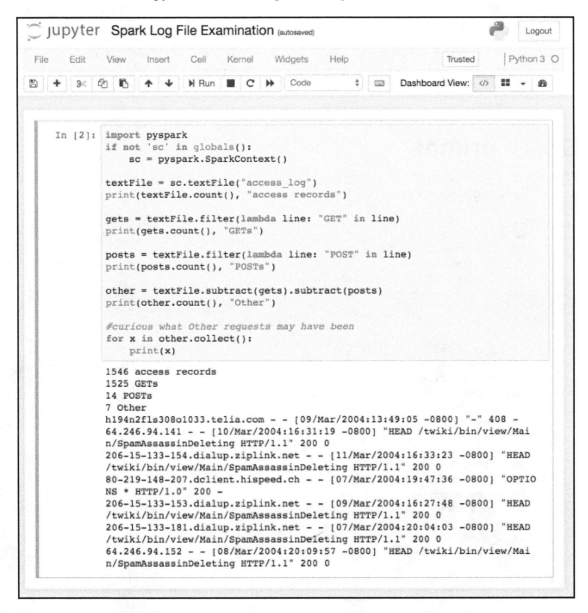

The text processing was not very fast (especially for so few records).

I liked being able to work with the data frames in such a way. There is something pleasing about being able to do basic algebra with sets in a programmatic way, without having to be concerned about edge cases.

By the way, a HEAD request works just like GET, but does not return the HTTP body. This allows a caller to determine what kind of response would have come back and respond appropriately.

Spark primes

We can run a series of numbers through a filter to determine whether each number is prime or not. We can use this script:

```
import pyspark
if not 'sc' in globals():
    sc = pyspark.SparkContext()
def is_it_prime(number):
    #make sure n is a positive integer
    number = abs(number)
    #simple tests
    if number < 2:
        return False
    #2 is special case
    if number == 2:
        return True
    #all other even numbers are not prime
    if not number & 1:
        return False
    #divisible into it's square root
    for x in range(3, int(number**0.5)+1, 2):
        if number % x == 0:
            return False
    #must be a prime
    return True

# pick a good range
numbers = sc.parallelize(range(100000))

# see how many primes are in that range
print(numbers.filter(is_it_prime).count())
```

The script generates numbers up to 100000.

We then loop over each of the numbers and pass it to our filter. If the filter returns `True`, we get a record. Then, we just count how many results we found.

Running this in Jupyter, we see the following:

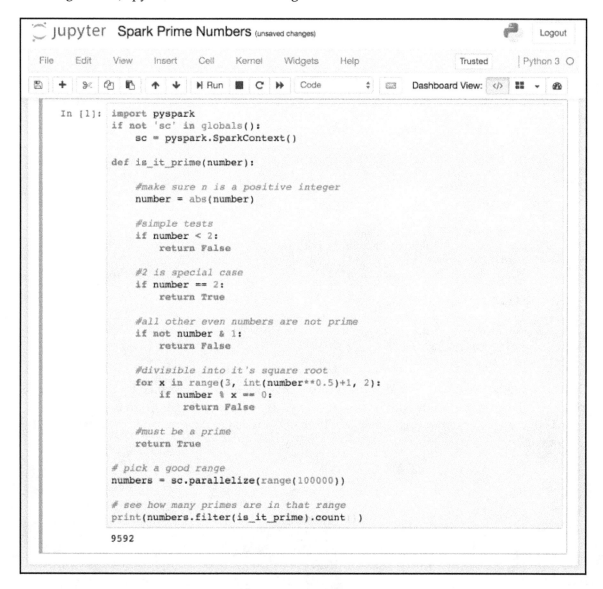

This was very fast. I was waiting and didn't notice that it went so quickly.

Spark text file analysis

In this example, we will look through a news article to determine some basic information from the article.

We will be using the following script against the 2600 raid news article from `https://www.newsitem.com`:

```python
import pyspark
if not 'sc' in globals():
    sc = pyspark.SparkContext()

#pull out sentences from article
sentences = sc.textFile('2600raid.txt') \
    .glom() \
    .map(lambda x: " ".join(x)) \
    .flatMap(lambda x: x.split("."))
print(sentences.count(),"sentences")

#find the bigrams in the sentences
bigrams = sentences.map(lambda x:x.split()) \
    .flatMap(lambda x: [((x[i],x[i+1]),1) for i in range(0, len(x)-1)])
print(bigrams.count(),"bigrams")

#find the (10) most common bigrams
frequent_bigrams = bigrams.reduceByKey(lambda x,y:x+y) \
    .map(lambda x:(x[1], x[0])) \
    .sortByKey(False)
frequent_bigrams.take(10)
```

The code reads in the article and splits it into `sentences`, as determined by the appearance of a period. From there, the code maps out the `bigrams` present. A bigram is a pair of words that appear next to each other. We then sort the list and print out the top ten most prevalent.

When we run this in a Notebook, we see these results:

```
jupyter   Spark Bigrams from News (autosaved)                          Logout

File    Edit    View    Insert    Cell    Kernel    Widgets    Help      Trusted    | Python 3 O

[save] + [cut] [copy] [paste] ↑ ↓ ▶ Run ■ C ▶▶ Code  ▼    [kbd]  Dashboard View: </>  ▦  ▾  ▩

In [1]:  import pyspark
         if not 'sc' in globals():
             sc = pyspark.SparkContext()

         #pull out sentences from article
         sentences = sc.textFile('2600raid.txt') \
             .glom() \
             .map(lambda x: " ".join(x)) \
             .flatMap(lambda x: x.split("."))
         print(sentences.count(),"sentences")

         #find the bigrams in the sentences
         bigrams = sentences.map(lambda x:x.split()) \
             .flatMap(lambda x: [((x[i],x[i+1]),1) for i in range(0, len(x)-1)])
         print(bigrams.count(),"bigrams")

         #find the (10) most common bigrams
         frequent_bigrams = bigrams.reduceByKey(lambda x,y:x+y) \
             .map(lambda x:(x[1], x[0])) \
             .sortByKey(False)
         frequent_bigrams.take(10)

         220 sentences
         3463 bigrams

Out[1]:  [(36, ('of', 'the')),
          (15, ('the', 'mall')),
          (12, ('to', 'the')),
          (12, ('on', 'the')),
          (12, ('At', 'this')),
          (11, ('the', 'guards')),
          (11, ('at', 'the')),
          (11, ('in', 'the')),
          (9, ('a', 'few')),
          (9, ('and', 'the'))]
```

I really had no idea what to expect from the output. It's curious that you can glean some insight into the article as `the` and `mall` appear 15 times and `the` and
`guards` appear 11 times—a raid must have occurred in a mall and included the security guards in some manner.

Spark evaluating history data

In this example, we combine the previous sections to look at some historical data and determine a number of useful attributes.

The historical data we are using is the guest list for the Jon Stewart television show. A typical record from the data looks as follows:

```
1999,actor,1/11/99,Acting,Michael J. Fox
```

This contains the year, the occupation of the guest, the date of appearance, a logical grouping of the occupations, and the name of the guest.

For our analysis, we will be looking at the number of appearances per year, the occupation that appears most frequently, and the personality who appears most frequently.

We will be using this script:

```python
#Spark Daily Show Guests
import pyspark
import csv
import operator
import itertools
import collections

if not 'sc' in globals():
 sc = pyspark.SparkContext()

years = {}
occupations = {}
guests = {}

#file header contains column descriptors:
#YEAR, GoogleKnowledge_Occupation, Show, Group, Raw_Guest_List

with open('daily_show_guests.csv', 'rt', errors = 'ignore') as csvfile:
 reader = csv.DictReader(csvfile)
 for row in reader:
 year = row['YEAR']
 if year in years:
 years[year] = years[year] + 1
 else:
 years[year] = 1

 occupation = row['GoogleKnowlege_Occupation']
 if occupation in occupations:
 occupations[occupation] = occupations[occupation] + 1
```

```
    else:
    occupations[occupation] = 1

    guest = row['Raw_Guest_List']
    if guest in guests:
    guests[guest] = guests[guest] + 1
    else:
    guests[guest] = 1

#sort for higher occurrence
syears = sorted(years.items(), key = operator.itemgetter(1), reverse =
True)
soccupations = sorted(occupations.items(), key = operator.itemgetter(1),
reverse = True)
sguests = sorted(guests.items(), key = operator.itemgetter(1), reverse =
True)

#print out top 5's
print(syears[:5])
print(soccupations[:5])
print(sguests[:5])
```

The script has a number of features:

- We are using several packages.
- It has the familiar context preamble.
- We start dictionaries for `years`, `occupations`, and `guests`. A dictionary contains `key` and `value`. For this use, the key will be the raw value from the CSV. The value will be the number of occurrences in the dataset.
- We open the file and start reading line by line, using a `reader` object. We are using ignore errors as there are a couple of nulls in the file.
- On each line, we take the value of interest (`year`, `occupation`, `name`).
- We see if the value is present in the appropriate dictionary.
- If it is there, increment the value (counter).
- Otherwise, initialize an entry in the dictionary.
- We then sort each of the dictionaries in reverse order of the number of appearances of the item.
- Finally, we display the top five values for each dictionary.

If we run this in a Notebook, we have the following output:

```
for row in reader:
    year = row['YEAR']
    if year in years:
        years[year] = years[year] + 1
    else:
        years[year] = 1

    occupation = row['GoogleKnowlege_Occupation']
    if occupation in occupations:
        occupations[occupation] = occupations[occupation] + 1
    else:
        occupations[occupation] = 1

    guest = row['Raw_Guest_List']
    if guest in guests:
        guests[guest] = guests[guest] + 1
    else:
        guests[guest] = 1

#sort for higher occurrence
syears = sorted(years.items(), key = operator.itemgetter(1), reverse = True
soccupations = sorted(occupations.items(), key = operator.itemgetter(1), re
sguests = sorted(guests.items(), key = operator.itemgetter(1), reverse = Tr

#print out top 5's
print(syears[:5])
print(soccupations[:5])
print(sguests[:5])

[('2000', 169), ('1999', 166), ('2003', 166), ('2013', 166), ('2010', 16
5)]
[('actor', 596), ('actress', 271), ('journalist', 180), ('author', 102),
('Journalist', 72)]
[('Fareed Zakaria', 19), ('Denis Leary', 17), ('Brian Williams', 16), ('P
aul Rudd', 13), ('Ricky Gervais', 13)]
```

We show the tail of the script and the preceding output.

There may be a smarter way to do all of this, but I am not aware of it.

The build-up of the accumulators is pretty standard, regardless of what language you are using. I think there is an opportunity to use a `map()` function here.

I really liked just trimming off the lists/arrays so easily instead of having to call a function.

The number of guests per year is very consistent. Actors are prevalent—probably the group of people of most interest to the audience. The guest list was a little surprising. The guests are mostly actors, but I think all have strong political directions.

Summary

In this chapter, we used Spark functionality via Python coding for Jupyter. First, we installed the Spark additions to Jupyter. We wrote an initial script that just read lines from a text file. We went further and determined the word counts in that file. We added sorting to the results. We wrote was a script to estimate pi. We evaluated web log files for anomalies. We determined a set of prime numbers, and we evaluated a text stream for certain characteristics.

Interactive Widgets

9

There is a mechanism for Jupyter to gather input from the user while the script is running. To do this, we put coding in the use of a widget or user interface control in the script. The widgets we will use in this chapter are defined at `http://ipywidgets.readthedocs.io/en/latest/`

For example, there are widgets for the following:

- **Text input**: The Notebook user enters a string that will be used later in the script.
- **Button clicks**: These present the user with multiple options by way of buttons. Then, depending on which button is selected (clicked on); your script can change direction according to the user.
- **Slider**: You can provide the user with a slider where the user can select a value within the range you specify, and then your script can use that value accordingly.
- **Toggle box and checkboxes**: These let the user select the different options of your script that they are interested in working with.
- **Progress bar**: A progress bar can be used to show how far along they are in a multi-step process.

In some cases, this can be wide open as the underlying *gather input from the user* is generally available. Therefore, you could make really interesting widgets that do not fit the standard user, input a control paradigm. For example, there is a widget allowing a user to click on a geographical map to discover data.

In this chapter, we will cover the following topics:

- Installing widgets
- Widget basics
- Interact widget
- Interactive widget
- Widgets package

Installing widgets

- The widgets package is an upgrade to the standard Jupyter installation. You can update the widgets package using this command:

```
pip install ipywidgets
```

- Once complete, you must then upgrade your Jupyter installation using this command:

```
jupyter nbextension enable --py widgetsnbextension
```

- And then you must use this command:

```
conda update jupyter_core jupyter_client
```

- We put together a basic example widget Notebook to make sure everything is working:

```
 #import our libraries
from ipywidgets import *
from IPython.display import display

#create a slider and message box
slider = widgets.FloatSlider()
message = widgets.Text(value = 'Hello World')

#add them to the container
container = widgets.Box(children = (slider, message))
container.layout.border = '1px black solid'

display(container)
```

- We end up with the following screenshot, where the container widget is displayed enclosing the slider and the message box:

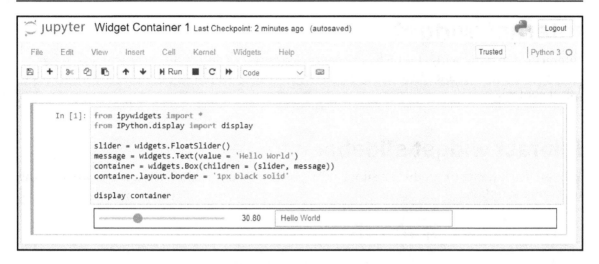

Widget basics

All widgets work the same, generally:

- You can create or define an instance of a widget.
- You can preset the attributes of a widget, such as its initial value, or a label to be displayed.
- Widgets can react to different inputs from a user. The inputs are gathered by a handler or Python function that is invoked when a user performs some action on a widget, for example, to call the handler if the user clicks on a button.
- The value of a widget can be used later in your script just as any other variable. For example, you can use a widget to determine how many circles to draw.

Interact widget

Interact is the basic widget which is appears to be used to derive all other widgets. It has variable arguments, and depending on the arguments, will portray a different kind of user input control.

Interact widget slidebar

We can use interact to produce a slidebar by passing in an extent. For example, we have the following script:

```
#imports
from ipywidgets import interact

# define a function to work with (cubes the number)
def myfunction(arg):
    return arg+1

#take values from slidebar and apply to function provided
interact(myfunction, arg=9);
```

 Note that the semicolon following the `interact` function call is required.

We have a script which does the following:

- References the package we want to use
- Defines a function that is called for every user input of a value
- Calls to interact, passing our handler and a range of values

When we run this script, we get a scrollbar that is modifiable by the user:

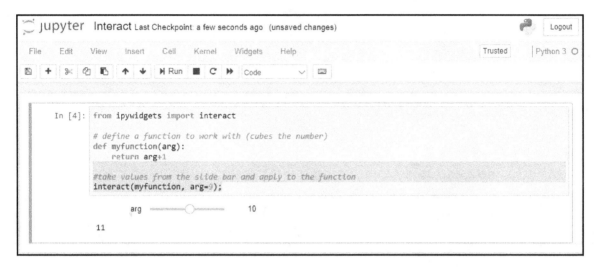

The user is able to slide the vertical bar over the range of values. The upper end is 27 and the lower end is -1 (assuming we could pass additional arguments to interact to set the range of values selectable). myfunction is called every time the value in the interact widget is changed and the result printed. Hence, we see 27 selected and the number 28 displayed (following the processing of myfunction - 27 + 1).

Interact widget checkbox

We can change the type of control generated, based on the arguments passed to interact. If we had the following script:

```
from ipywidgets import interact
def myfunction(x):
    return x
# the second argument defines which of the interact widgets to use
interact(myfunction, x=False);
```

We are doing the same steps as before; however, the value passed is False (but it could also be True). The interact function examines the argument passed, determines it is a Boolean value, and presents the appropriate control for a Boolean: a checkbox.

If we run the preceding script in a Notebook, we get a display like the following:

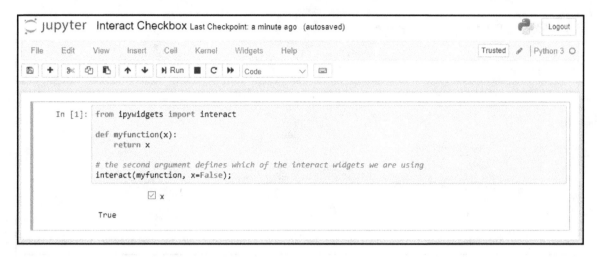

Interact widget textbox

We can generate a text input control, again by passing in different arguments to interact. For example, see the following script:

```
from ipywidgets import interact
def myfunction(x):
    return x
#since argument is a text string, interact generates a textbox
control for it
interact(myfunction, x= "Hello World");
```

This produces a text input control with the initial value of `Hello World`:

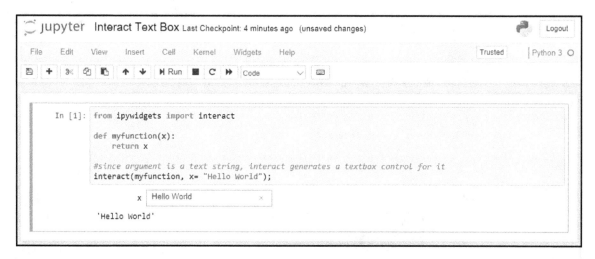

Interact dropdown

We can also use the interact function to produce a drop-down listbox for the user to select from. In the following script we produce a dropdown with two choices:

```
from ipywidgets import interact
def myfunction(x):
    return x
interact(myfunction, x=['red','green']);
```

This script will do the following:

- Pull in the interact reference,
- Define a function that will be called whenever the user changes the value of the control
- Calls the `interact` function with a set of values, which interact will interpret to mean create a dropdown for the user.

If we run this script in a Notebook, we get the following display:

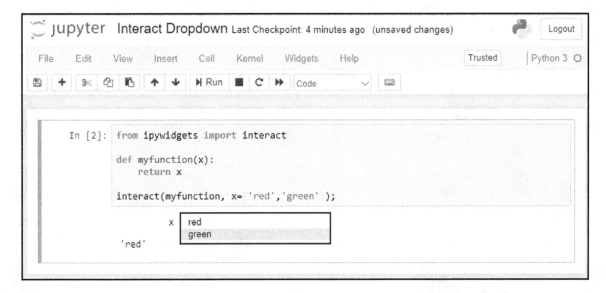

The preceding screenshot shows us that the value printed at the bottom will change according to what is selected in the dropdown by the user.

Interactive widget

There is also an interactive widget. The interactive widget works like the interact widget, but does not display the user input control until called upon directly by the script. This would be useful if you had some calculations that had to be performed for the parameters of the widget display, or even if you wanted to decide whether you needed a control at runtime.

For example, we could have a script (similar to the previous script) as follows:

```
from ipywidgets import interactive
from IPython.display import display
def myfunction(x):
return x
w = interactive(myfunction, x= "Hello World ");
display(w)
```

We have made a couple of changes to the script:

- We are referencing the interactive widget
- The interactive function returns a widget, rather than immediately displaying a value
- We must script the display of the widget ourselves

If we run the following script, to the user it looks very similar:

Widgets

There is another package of widgets, called `widgets`, that has all of the standard controls you might want to use, with many optional parameters available to customize your display.

The progress bar widget

One of the widgets available in this package displays a progress bar to the user. We could have the following script:

```
import ipywidgets as widgets
widgets.FloatProgress(
```

```
        value=45,
        min=0,
        max=100,
        step=5,
        description='Percent:',
    )
```

The preceding script would display our progress bar as follows:

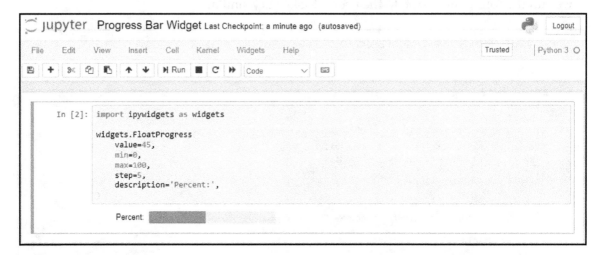

We see a progress bar that looks to be 45%.

The listbox widget

We could also use the listbox widget, called a Dropdown, as in the following script:

```
import ipywidgets as widgets
from IPython.display import display
w = widgets.Dropdown(
    options={'Pen': 7732, 'Pencil': 102, 'Pad': 33331},
    description='Item:',
)
display(w)
w.value
```

This script will display a listbox to the user with the values Pen, Pencil, and Pad. When the user selects one of the values, the associated value is returned in the w variable, which we display as in the following screenshot:

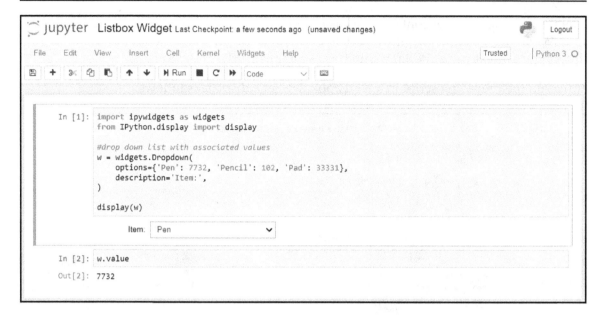

Therefore, we see the inventory value associated with Pen.

The text widget

The `text` widget gathers a text string from the user for reuse elsewhere in your script. A text widget has a description (label) and a value (entered by the user or preset in your script).

In this example, we will just gather a text string and display it on the screen as part of the output for the script. We will use the following script:

```
from ipywidgets import widgets
from IPython.display import display
#take the text from the box, define variable for handler
text = widgets.Text()
#display it
display(text)
def handle_submit(sender):
    print(text.value)
#when we hit return in the textbox call upon the handler
text.on_submit(handle_submit)
```

The Python package that contains the basic widgets is `ipywidgets`, so we need to reference that. We define a handler for the text field that will be called when the user hits submit after entering their text value. When we do this, we are using the `text` widget.

If we run the preceding script in a Notebook, we get a display as follows:

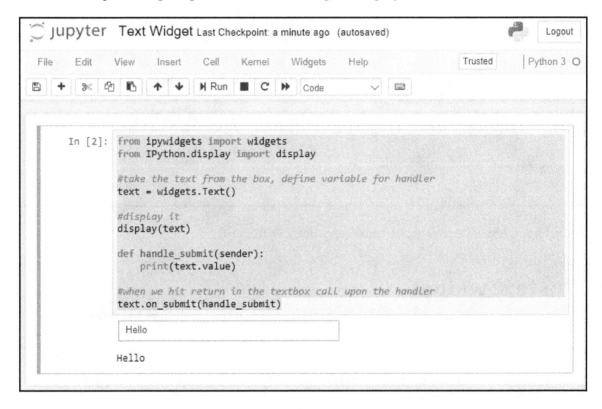

We should point out some of the highlights of this page:

- The ordering of the elements of the page is important. The text field referenced by the handler must be defined before the reference.
- When invoking a widget, the widget automatically looks for any handlers that might be associated with the script. In this case, we have a `submit` handler. There are many other handlers available. `text.on_submit` assigns the handler to the widget.
- Otherwise, we have a standard Python Notebook.

- If we run the script (**Cell** | **Run All**), we get the preceding screenshot (waiting for us to enter a value in the text box):
- So, our script has set up a widget for gathering input from the user and later done something with that value. (We are just displaying it here, but we could use the input for further processing.)

The button widget

Similarly, we can use a Button widget in our script, as can be seen in the following example:

```
from ipywidgets import widgets
from IPython.display import display

button = widgets.Button(description="Submit");
display(button)

def on_button_clicked(widget):
    print("Clicked Button:" + widget.description);
button.on_click(on_button_clicked);
```

This script does the following:

- References the features we want to use from the widgets packages.
- Creates our button.
- Defines a handler for the click event on a button. The handler receives the Button object that was clicked upon (the widget).
- The handler displays information about the button clicked on (you can imagine if we had several buttons in a display, we would want to determine which button was clicked).
- Lastly, we assign the defined handler to the Button object we created.

 Note the indentation of the coding for the handler; this is the standard Python style that must be followed.

If we run the preceding script in a Notebook, we get a display like the following screenshot:

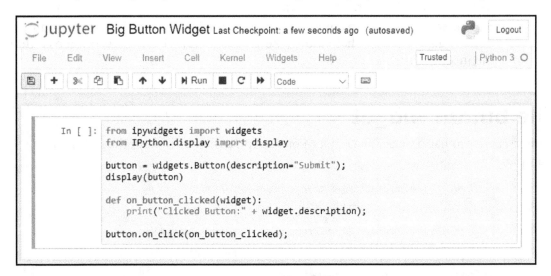

Note the Submit button at the bottom of the following image. You could change other attributes of the button, such as its position, size, color, and many more.

If we then click on the Submit button, we get the following display where our message about the button being clicked is displayed:

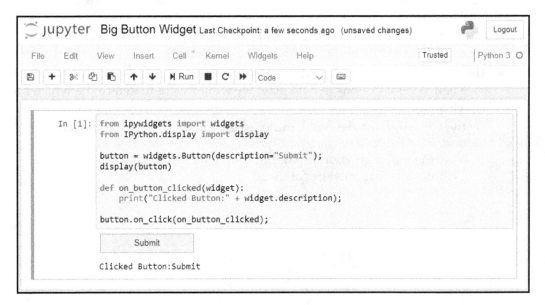

Widget properties

All widgets controls have a set of properties for your display that can be adjusted as needed. You can see the list of properties by taking an instance of a control and running the `control.keys` command in a Notebook, as can be seen in the following example:

```
from ipywidgets import *
w = IntSlider()
w.keys
```

This script pulls in a blanket reference to all of the controls available in widgets. We then create an `IntSlider` instance and display the list of properties that we can adjust. So, we end up with a display like the following screenshot:

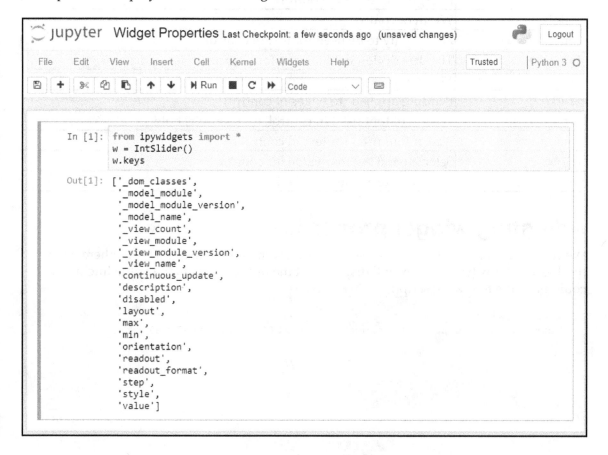

As you can see, the list is extensive:

Property	Description
orientation	Whether left align, right align or justified
color	Color of font
height	Height of control
disabled	Whether control is disabled or not
visible	Is the control displayed?
font_style	Style of font, for example, italic
min	Minimum value (used in range list)
background_color	Background color of control
width	Width of control
font_family	Font family to be used for text in control
description	The description field is used for documentation purposes
max	Maximum value (of range)
padding	Padding applied (to edges of control)
font_weight	Weight of font used in control, for example, bold
font_size	Size of font used for text in control
value	Selected and entered value for control
margin	Margin to use when displaying control

Adjusting widget properties

We could adjust any of these in our scripts using something like this script, where we disable a text box (the text box will display, but the user cannot enter a value into it). We could have the following script:

```
from ipywidgets import *
Text(value='You can not change this text!', disabled=True)
```

This is the screenshot of the preceding code:

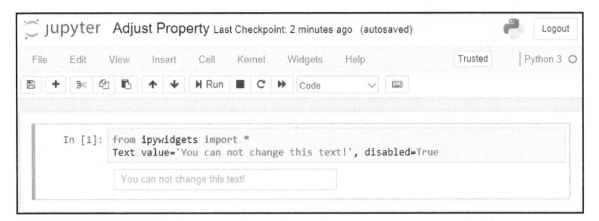

When a field is disabled, the textbox is grayed out. When the user hovers the cursor over the field, they get a red circle with a slash through it, which means it cannot be changed.

Adjusting properties

All of the properties shown previously are accessible to read and write. We can show this transition with a small script, as follows:

```
from ipywidgets import *
w = IntSlider()
original = w.value
w.value = 5
original, w.value
```

The script creates a slider, retrieves its current value, changes the value to 5, and then displays the original and current value of the control.

If we were to run the preceding script in a Notebook, we would see the following expected results:

Widget events

All of the controls work by reacting to user actions, either with a mouse or keyboard. The default actions for controls are built into the software, but you can add your own handling of events (user actions).

We have seen this kind of event handling previously (for example, in the section on the slidebar, a function is called whenever the slider value is changed by the user). But, let's explore it in a little more depth.

We could have the following script:

```
from ipywidgets import widgets
from IPython.display import display
button = widgets.Button(description="Click Me!")
display(button)
def on_button_clicked(b):
    print("Button clicked.")
button.on_click(on_button_clicked)
```

This script does the following:

- Creates a button.
- Displays the button to the user.
- Defines a handler click event. It prints a message that you clicked on the screen. You can have any Python statements you want in the handler.
- Lastly, we associated the click handler with the button we created. So, now when the user clicks on our button, the handler is called and the `Button clicked` message is displayed on screen (as shown in the following screenshot):

If we run the preceding script in a Notebook and click on the button a few times, we get the following display:

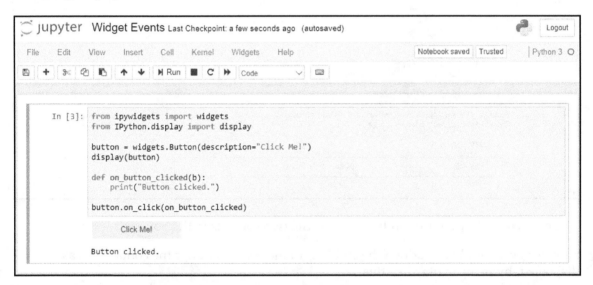

Widget containers

You can also assemble containers of widgets directly, using Python syntax, by passing the child elements in the constructor. For example, we could have the following script:

```
#import our libraries
from ipywidgets import *
from IPython.display import display

#create a slider and message box
slider = widgets.FloatSlider()
message = widgets.Text(value = 'Hello World')
```

```
#add them to the container
container = widgets.Box(children = (slider, message))
container.layout.border = '1px black solid'

display(container)
```

The preceding script shows that we are creating a container (which is a `Box` widget), in which we are specifying the child, contained controls. The call to display the container will iteratively display the child elements as well. So, we end up with a display like the following screenshot:

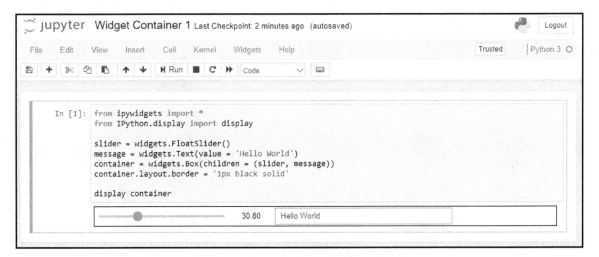

You can see the border around the box and the two controls in the box.

Similarly, we could have added the children to the container after the container was displayed, by using syntax like this:

```
from ipywidgets import *
from IPython.display import display
container = widgets.Box()
container.layout.border = '1px black solid'
display(container)
slider = widgets.FloatSlider()
message = widgets.Text(value='Hello World')
container.children=[slider, message]
```

When we add the child to the container, the container repaints, which will cause the repainting of any children.

If we run this script in another Notebook, we get a very similar result to the previous example, with a display like the following screenshot:

Summary

In this chapter, we added widgets to our Jupyter installation and we used the interact and interactive widgets to produce a variety of user input controls. We then looked at the widgets package in depth to investigate some of the user controls available, properties available in the containers, events that are available emitting from the controls, and how to build containers for the controls.

In the next chapter, we will look into sharing Notebooks and converting them into different formats.

10
Sharing and Converting Jupyter Notebooks

Once you have developed your Notebook, you will want to share it with others. There is a typical mechanism available for sharing that we will cover in this chapter—placing your Notebook on an accessible server on the internet.

When you provide a Notebook to another person, they may need the Notebook in a different format, given their system requirements. We will also cover some mechanisms available for providing your Notebook to others in a different format.

In this chapter, we will cover the following topics:

- Sharing Notebooks
- Converting Notebooks

Sharing Notebooks

The typical mechanism for sharing Notebooks is to provide your Notebook on a website. A website is running on a server or allocated machine space. The server takes care of all the book-keeping involved in running a website, such as keeping track of multiple users and logging people on and off.

In order for the Notebook to be of use though, the website must have Notebook logic installed. A typical website knows how to deliver content as HTML given some source files. The most basic form is pure HTML, where every page you access on the website corresponds exactly to one HTML file on the web server. Other languages could be used to develop the website (such as Java or PHP), so then the server needs to know how to access the HTML it needs from those source files. In our context, the server needs to know how to access your Notebook in order to deliver HTML to users.

Even when Notebooks are just running on your local machine, they are running in a browser that is accessing your local machine (server) instead of the internet—so the web, HTML, and internet access have already been provided.

If a Notebook is on a genuine website, it is available to everyone who can access that website—irrespective of whether the server is running on your machine in an office environment accessible over the local area network, or it is available to all users over the internet.

You can always add security around the website so that the only people who can use your Notebook are those given access by you. Security mechanisms depend on the type of web server software involved.

Sharing Notebooks on a Notebook server

Built into the Jupyter process is the ability to expose a Notebook as its own web server. Assuming the server is a machine accessible by other users, you can configure Jupyter to run on that server. You must provide the configuration information to Jupyter so that it knows how to proceed. The command to generate a configuration file for your Jupyter installation is shown in the following example:

 Note that, as we are running in Anaconda, I am running the command from that directory:

```
C:\Users\Dan\Anaconda3\Scripts>jupyter notebook --generate-config
```

The default `config` is written
to: `C:\Users\Dan\.jupyter\jupyter_notebook_config.py`.

This command will generate a `jupyter_notebook_config.py` file in your `~./jupyter` directory. For Microsoft users, that directory is a subdirectory of your home user directory.

The configuration file contains the settings that you can use to expose your Notebook as a server:

```
c.NotebookApp.certfile = u'/path/to/your/cert/cert.pem'
c.NotebookApp.keyfile = u'/ path/to/your/cert/key.key'
c.NotebookApp.ip = '*'
c.NotebookApp.password = u'hashed-password'
c.NotebookApp.open_browser = False
c.NotebookApp.port = 8888
```

The settings in the file are explained in the following table:

Setting	Description
c.NotebookApp.certfile	This is the path to the location of the certificate for your site. If you have an SSL certificate, you would need to change the setting to the location of the file. It may not be a .PEM extension file. There are several SSL certificate formats.
c.NotebookApp.keyfile	This is the path to the location of the key to access the certificate for your site. Rather than specify the key to your certificate, you would have stored the key in a file. So, if you want to apply an SSL certificate to your Notebook, you would specify the file location. The key is normally a very large hexadecimal number. Hence, it is stored in its own file. Also, storing it in a file offers additional protection, since the directory where keys are stored on a machine usually has limited access.
c.NotebookApp.ip	IP address of the machine. Use the wildcard * to open it to everyone. Here, we are specifying the IP address of the machine where the Notebook website is accessed.
c.NotebookApp.password	Hashed password—the password will have to be provided by users accessing your Notebook in response to a standard login challenge.
c.NotebookApp.open_browser	True/False—does starting the Notebook server open a browser window?
c.NotebookApp.port	Port to access your server; it should be open to the machine.

Every website is addressed at the lower levels by an IP address. An IP address is a four-part number that identifies the locale of the server involved. An IP address might look like `172.32.88.7`. Web browsers, in concert with internet software, know how to use the IP address to locate the server of interest. The set of software also knows how to translate the URL you mentioned in your browser, such as `http://www.microsoft.com`, into an IP address.

Once you have changed the settings appropriately, you should be able to point a browser at the URL configured and access your Notebook. The URL would be the concatenation of either HTTP or HTTPS (depending on whether you applied an SSL certificate), the IP address, and the port, for example, `HTTPS://123.45.56.9:8888`.

Sharing encrypted Notebooks on a Notebook server

Two of the settings shown earlier can be used if you have an SSL certificate to apply. Without the SSL certificate, the password (seen previously) and all other interactions will be transmitted from the user's browser to the server in clear text. If you are dealing with sensitive information in your Notebook, you should obtain an SSL certificate and make the appropriate setting changes for your server.

If you need more security regarding access to your Notebook, the next step would be to provide an SSL certificate (placed on your machine, with the path provided in the configuration file). There are a number of companies that provide SSL certificates. The cheapest, at the time of writing, is *Let's encrypt*, which will provide a low-level SSL certificate for free. (There are gradations of SSL certificates that are not free.)

Again, once you have set the preceding settings with regard to your certificate, you should be able to access your Notebook server using the `HTTPS://` prefix, knowing that all the transmissions between the user's browser and the Notebook server are encrypted and therefore secure.

Sharing Notebooks on a web server

In order to add your Notebook to an existing web server, you need to take the preceding steps and add a little more information to the Notebook configuration file, as in the following example:

```
c.NotebookApp.tornado_settings = {
'headers': {
'Content-Security-Policy': "frame-ancestors 'https://yourwebsite.com'
'self' "
}
}
```

Here, you replace yourwebsite.com with the URL of your website.

Once complete, you can access the Notebook through your website.

Sharing Notebooks through Docker

Docker is an open, lightweight container for distributing software. A typical Docker instance has an entire web server and a specific web application running on a port in a machine. The specifics about the software running in a Docker instance are governed by the Dockerfile. This file provides commands to the Docker environment as to which components to use to configure this instance. A sample Dockerfile for a Jupyter implementation would look like this:

```
ENV TINI_VERSION v0.6.0

ADD https://github.com/krallin/tini/releases/download/${TINI_VERSION}/tini
/usr/bin/tini
RUN chmod +x /usr/bin/tini
ENTRYPOINT ["/usr/bin/tini", "--"]
EXPOSE 8888
CMD ["jupyter", "Notebook", "--port=8888", "--no-browser", "--ip=0.0.0.0"]
```

The following is a discussion of each of the Dockerfile commands:

- The `ENV` command tells Docker to use a specialized operating system. This is necessary to overcome a deficiency in Jupyter that keeps obtaining and releasing resources from your machine.
- The `ADD` command tells Docker where the `tini` code is located.
- The `RUN` command changes access rights to the `tini` directory.
- The `ENTRYPOINT` command tells Docker what to use as the operating system of the Docker instance.
- The `EXPOSE` command tells Docker what port to expose your Notebook on.
- The `CMD` command tells Docker what commands to run (once the environment is set up). The `CMD` arguments are telling as you see the familiar `jupyter Notebook` command that you use to start Jupyter on your machine.

Once the Docker instance is deployed to your Docker machine, you can access the Docker instance on the machine at the port specified (`8888`), such as, `http://machinename.com:8888`.

Sharing Notebooks on a public server

Currently, there is one hosting company that allows you to host your Notebooks for free: GitHub. GitHub is the standard web provider of source control (Git source control) systems. Source control is used to maintain historical versions of your files to allow you to retrace your steps.

GitHub's implementation includes all of the tools that you need to use in your Notebook already installed on the server. For example, in previous chapters, to use R programming in your Notebook, you would have had to install the R toolset on your machine. GitHub has already done all of these steps.

Go to the `https://github.com/` website and sign up for a free website.

Once logged in, you are provided with a website that can be added to. If you have development tools to use (`git` push commands are programmer commands to store files on a Git server), you can do that or simply drag and drop you Notebook (`ipynb`) file onto your GitHub website.

I created an account there, with a `notebooks` directory, and placed one of the `notebooks` on that site. My GitHub site looks as follows:

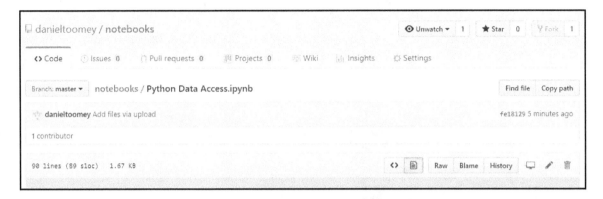

You can see the `Python Data Access.ipynb` file near the top of the screen.

If you click on that `notebooks` file, you see the expected Notebook up and running in your browser.

 Note: There is currently a problem running `notebooks` on GitHub. It was working. I expect something will get fixed to re-enable Jupyter.

If you look back at the previous chapter, you can see the same display (minus the GitHub adornments).

This Notebook is directly accessible by others using this URL: `https://github.com/danieltoomey/notebooks/blob/master/Python%20Data%20Access.ipynb`. So, you can provide your Notebook on GitHub to other users and just hand them the URL.

 When you are logged in to GitHub, the display will look slightly different as you will have more control over the GitHub content.

Converting Notebooks

The standard tool for converting Notebooks to other formats is the use of the `nbconvert` utility. It is built into your Jupyter installation. You can access the tool directly in the user interface for your Notebook. If you open a Notebook and select the Jupyter **File** menu item, you will see several options for **Download as**:

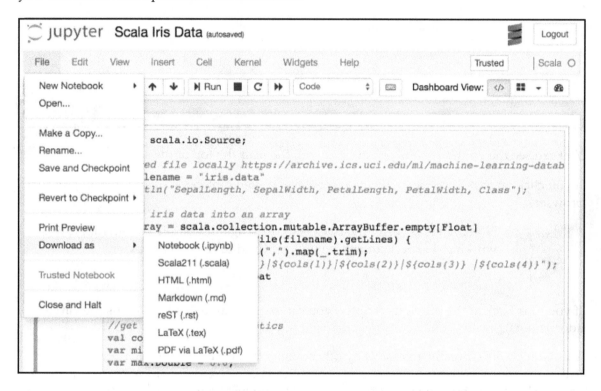

The choices are:

Format type	File extension
Notebook	`.ipynb`
Scala211	`.scala`
HTML	`.html`
Markdown	`.md`
reST	`.rst`
LaTeX	`.tex`
PDF via LaTeX	`.pdf`

Note: Since we are working with a Scala Notebook, that is the language choice provided on the second choice. If we had a Notebook in another language, that other language would be the choice.

For these examples, if we take a Notebook from a previous chapter, the Jupyter Notebook looks like this:

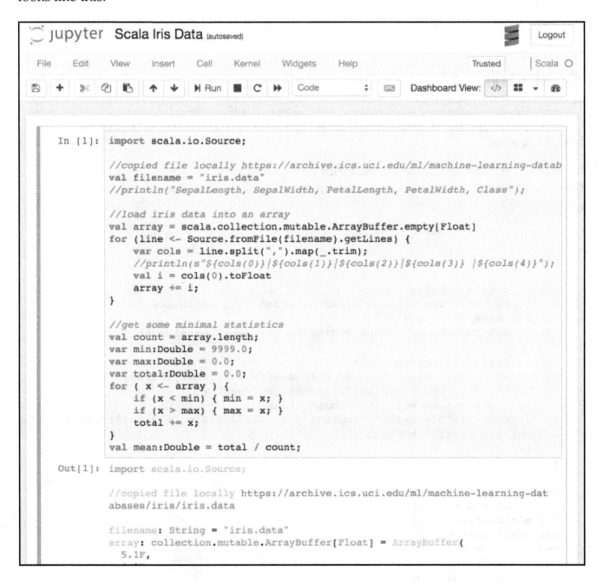

Notebook format

The Notebook format (`.ipynb`) is the native format for your Notebook. We have looked in this file in earlier chapters to see what Jupyter is storing in your Notebook.

You would use the Notebook format if you wanted to give another user complete access to your Notebook, since they would run your Notebook from their system.

You may also want to do this to save your Notebook in another medium.

Scala format

The Scala format (`.scala`) corresponds to the Scala implementation of your Notebook. If you use JavaScript as the language for your Notebook, this is a direct export of the Notebook page.

If you used another language for the script of the Notebook, Python, for example, then the **Download as** option would change appropriately, for example, **Download as | Python** (`.py`).

Using our example, as expected the Scala format is equivalent to the Jupyter display:

```scala
import scala.io.Source;

//copied file locally
https://archive.ics.uci.edu/ml/machine-learning-databases/iris/iris.data
val filename = "iris.data"
//println("SepalLength, SepalWidth, PetalLength, PetalWidth, Class");

//load iris data into an array
val array = scala.collection.mutable.ArrayBuffer.empty[Float]
for (line <- Source.fromFile(filename).getLines) {
    var cols = line.split(",").map(_.trim);
    //println(s"${cols(0)}|${cols(1)}|${cols(2)}|${cols(3)} |${cols(4)}");
    val i = cols(0).toFloat
    array += i;
}

//get some minimal statistics
val count = array.length;
var min:Double = 9999.0;
var max:Double = 0.0;
var total:Double = 0.0;
for ( x <- array ) {
    if (x < min) { min = x; }
```

```
        if (x > max) { max = x; }
        total += x;
    }
    val mean:Double = total / count;
```

With the Scala format, you can run the script directly using the Scala interpreter. Under macOS, there is the `scala` command. Similar tools exist for Windows machines.

Also, for other scripting languages, you should be able to run the script in the appropriate interpreter.

If we run this Scala file from a command-line window, we see the expected results:

```
scala> //copied file locally https://archive.ics.uci.edu/ml/machine-learning-databases/iris/iris.data

scala> val filename = "iris.data"
filename: String = iris.data

scala> //println("SepalLength, SepalWidth, PetalLength, PetalWidth, Class");

scala>

scala> //load iris data into an array

scala> val array = scala.collection.mutable.ArrayBuffer.empty[Float]
array: scala.collection.mutable.ArrayBuffer[Float] = ArrayBuffer()

scala> for (line <- Source.fromFile(filename).getLines) {
     |       var cols = line.split(",").map(_.trim);
     |       //println(s"${cols(0)}|${cols(1)}|${cols(2)}|${cols(3)} |${cols(4)}");
     |       val i = cols(0).toFloat
     |       array += i;
     | }

scala>

scala> //get some minimal statistics

scala> val count = array.length;
count: Int = 150

scala> var min:Double = 9999.0;
min: Double = 9999.0

scala> var max:Double = 0.0;
max: Double = 0.0

scala> var total:Double = 0.0;
total: Double = 0.0

scala> for ( x <- array ) {
     |       if (x < min) { min = x; }
     |       if (x > max) { max = x; }
     |       total += x;
     | }

scala> val mean:Double = total / count;
mean: Double = 5.843333326975505
```

HTML format

HTML (.html) format corresponds to the HTML needed to display the page in a web browser as it appears in your Notebook. The generated HTML does not have any coding logic; it only has the HTML necessary to display a similar page.

The HTML format would be useful to convey to another user the results of your Notebook. You may want to do this if you wanted to email your Notebook to another user (where the raw HTML would be transported and viewable in an email client application).

HTML is also useful if you have a web service available where you can insert new pages. If the web server does not have support for Jupyter files (refer to the first section of this chapter), HTML may be your only choice. Similarly, you may not want to hand over your source Jupyter Notebook (.ipynb) file, even if the web server does support Jupyter.

The exported HTML format looks like this in a browser:

```
In [1]:   import scala.io.Source;

          //copied file locally https://archive.ics.uci.edu/ml/machine-learning-databases/iris/iris.data
          val filename = "iris.data"
          //println("SepalLength, SepalWidth, PetalLength, PetalWidth, Class");

          //load iris data into an array
          val array = scala.collection.mutable.ArrayBuffer.empty[Float]
          for (line <- Source.fromFile(filename).getLines) {
              var cols = line.split(",").map(_.trim);
              //println(s"${cols(0)}|${cols(1)}|${cols(2)}|${cols(3)} |${cols(4)}");
              val i = cols(0).toFloat
              array += i;
          }

          //get some minimal statistics
          val count = array.length;
          var min:Double = 9999.0;
          var max:Double = 0.0;
          var total:Double = 0.0;
          for ( x <- array ) {
              if (x < min) { min = x; }
              if (x > max) { max = x; }
              total += x;
          }
          val mean:Double = total / count;

Out[1]:   import scala.io.Source;

          //copied file locally https://archive.ics.uci.edu/ml/machine-learning-databases/iris/iris.data

          filename: String = "iris.data"
          array: collection.mutable.ArrayBuffer[Float] = ArrayBuffer(
              5.1F,
              4.9F,
```

Notice that none of the Jupyter heading information is displayed or available. Otherwise, it looks the same as the Jupyter display.

Markdown format

The markdown (.md) format is a looser version of **Hyptertext Markdown Language (HTML)**. .md files can be used by some tools. It is normally used as the format of README files for software distributions (where the client's display capabilities may be very limited).

For example, the markdown format of the same Notebook is shown as follows:

```scala211
import scala.io.Source;

//copied file locally
https://archive.ics.uci.edu/ml/machine-learning-databases/iris/iris.data
val filename = "iris.data"
//println("SepalLength, SepalWidth, PetalLength, PetalWidth, Class");

//load iris data into an array
val array = scala.collection.mutable.ArrayBuffer.empty[Float]
for (line <- Source.fromFile(filename).getLines) {
    var cols = line.split(",").map(_.trim);
    //println(s"${cols(0)}|${cols(1)}|${cols(2)}|${cols(3)} |${cols(4)}");
    val i = cols(0).toFloat
    array += i;
}

//get some minimal statistics
val count = array.length;
var min:Double = 9999.0;
var max:Double = 0.0;
var total:Double = 0.0;
for ( x <- array ) {
    if (x < min) { min = x; }
    if (x > max) { max = x; }
    total += x;
}
val mean:Double = total / count;
```

Obviously, the Markdown format is a very low-level display. There are only minor text markings that help the reader determine the different formatting in use. I used the Atom editor to see what this looks like when interpreted:

```
Scala Iris Data.md
1
2
3    ```scala211
4    import scala.io.Source;
5
6    //copied file locally https://archive.ics.uci.edu/ml/machine-learning-databases/iris/iris.data
7    val filename = "iris.data"
8    //println("SepalLength, SepalWidth, PetalLength, PetalWidth, Class");
9
10   //load iris data into an array
11   val array = scala.collection.mutable.ArrayBuffer.empty[Float]
12   for (line <- Source.fromFile(filename).getLines) {
13       var cols = line.split(",").map(_.trim);
14       //println(s"${cols(0)}|${cols(1)}|${cols(2)}|${cols(3)} |${cols(4)}");
15       val i = cols(0).toFloat
16       array += i;
17   }
18
19   //get some minimal statistics
20   val count = array.length;
21   var min:Double = 9999.0;
22   var max:Double = 0.0;
23   var total:Double = 0.0;
24   for ( x <- array ) {
25       if (x < min) { min = x; }
26       if (x > max) { max = x; }
27       total += x;
28   }
29   val mean:Double = total / count;
30   ```
31
32
33
34
35       [32mimport [39m[36mscala.io.Source;
```

Again, a very clean display, still resembling the Jupyter Notebook display, with some odd coding at the bottom of the file and lots of escape characters. Not sure why.

Restructured text format

The restructured text format (`.rst`) is a simple, plain text markup language that is sometimes used for programming documentation. It looks very similar to the `.md` format discussed earlier.

For example, the RST format for the example page looks like this:

```scala
.. code:: scala

    import scala.io.Source;
    //copied file locally
https://archive.ics.uci.edu/ml/machine-learning-databases/iris/iris.data
    val filename = "iris.data"
    //println("SepalLength, SepalWidth, PetalLength, PetalWidth, Class");
    //load iris data into an array
    val array = scala.collection.mutable.ArrayBuffer.empty[Float]
    for (line <- Source.fromFile(filename).getLines) {
        var cols = line.split(",").map(_.trim);
        //println(s"${cols(0)}|${cols(1)}|${cols(2)}|${cols(3)}
|${cols(4)}");
        val i = cols(0).toFloat
        array += i;
    }
    //get some minimal statistics
    val count = array.length;
    var min:Double = 9999.0;
    var max:Double = 0.0;
    var total:Double = 0.0;
    for ( x <- array ) {
        if (x < min) { min = x; }
        if (x > max) { max = x; }
        total += x;
    }
    val mean:Double = total / count;

.. parsed-literal::
...
```

As you can see, it is similar to the markdown in the previous example; the code is roughly broken up into chunks.

Using Atom to display the .rst file results in this:

```
import scala.io.Source;

//copied file locally https://archive.ics.uci.edu/ml/machine-learning-databases/iris/iris.data
val filename = "iris.data"
//println("SepalLength, SepalWidth, PetalLength, PetalWidth, Class");

//load iris data into an array
val array = scala.collection.mutable.ArrayBuffer.empty[Float]
for (line <- Source.fromFile(filename).getLines) {
    var cols = line.split(",").map(_.trim);
    //println(s"${cols(0)}|${cols(1)}|${cols(2)}|${cols(3)} |${cols(4)}");
    val i = cols(0).toFloat
    array += i;
}

//get some minimal statistics
val count = array.length;
var min:Double = 9999.0;
var max:Double = 0.0;
var total:Double = 0.0;
for ( x <- array ) {
    if (x < min) { min = x; }
    if (x > max) { max = x; }
    total += x;
}
val mean:Double = total / count;




.. parsed-literal::
```

The .rst display is not as nice as some of the others.

LaTeX format

LaTeX is a typesetting format from the late 1970s. It is still used on many Unix-dervied systems for producing manuals and the like.

Jupyter uses the LaTeX package to export the image of the Notebook to a PDF file. You would have to install this package on your machine in order for this to work. On macOS, this involves the following:

- Install LaTeX – there are separate installations for Windows and macOS.
- On macOS, I tried using MacTeX. You must use Safari to download the package. It complained about bad formats and I had to retry several times.
- On Windows, I tried using TeXLive. It attempts to download hundreds of fonts.
- The following (macOS) commands are used for fonts:
  - `sudo tlmgr install adjustbox`
  - `sudo tlmgr install collection-fontsrecommended`

 Be aware that this install was pretty cumbersome. I had already installed the full LaTeX, then another note said to install a mini version of LaTeX, and then it was tricky to install the fonts. I have very little confidence that these steps will work correctly on a Windows machine.

If you do not have the full set of packages needed, when you try to download the PDF file, a new screen will open in your Notebook and a long error message will be displayed showing what piece is missing.

You can download a text file reader for the particular operating system you are using to work with. I downloaded MacTeX for my macOS.

 Note: you will need a Tex interpreter in order to perform the next type of download as a PDF, since it uses Tex as the basis for developing the PDF file.

PDF format

The PDF (`.pdf`) format is a well-known display format used for many purposes. PDF is a good format for conveying unmodifiable content to other users. The other users will not be able to modify the results in any way, but they will be able to see and understand your logic.

PDF generation is dependent upon LaTeX being installed correctly. I was not able to get this running this time. I have successfully installed it on Windows and Mac in prior versions.

Summary

In this chapter, we shared Notebooks on a Notebook server. We added a Notebook to our web server, and we distributed a Notebook using GitHub. We also looked into converting our Notebooks into different formats, such as HTML and PDF.

In the next chapter, we will look into allowing multiple users to interact with a Notebook simultaneously.

Multiuser Jupyter Notebooks

11

Jupyter Notebooks have the inherent ability to be modifiable by users when the user enters data or makes a selection. However, there is an issue with the standard implementation of the Notebook server software that does not account for more than one person working with a Notebook at the same time. The Notebook server software is the underlying Jupyter software that displays the page and interacts with the user—it follows the directions in your Notebook for display and interaction.

A Notebook server, really a specialized internet web server, typically creates a new path or thread of execution for each user to allow for multiple users. A problem arises when a lower-level subroutine, used for all instances, does not properly account for multiple users where each has their own set of data.

 Note that some of the coding/installs in this chapter may not work in a Windows environment.

In this chapter we will explore the following:

- Give an example of the issue that occurs when multiple users access the same Notebook in a standard Jupyter installation
- Use a new version of Jupyter, JupyterHub, that was built by extending Jupyter to specifically address the multiple user problem
- Also use Docker, a tool to allow for multiple instances of any software, in order to address the issue

A sample interactive Notebook

For this chapter, we will use a simple Notebook that asks the user for some information, and displays certain information.

For example, we could have a script like this (taken from the previous `Chapter 9`, *Interactive Widgets*):

```
from ipywidgets import interact
def myfunction(x):
    return x
interact(myfunction, x= "Hello World ");
```

The script presents a textbox to the user with the original value of the box containing the `Hello World` string. As the user interacts with the input field and changes the value, then the value of the x variable in the script changes accordingly and is displayed on screen. For example, I have changed the value to letter `A`:

We can see the multiuser problem: if we just open the same page in another browser window (copy the URL, open a new browser window, paste in the URL, and hit the *Enter* key), we get the exact same display. The new window should have started with a new script, just prompting you with the default `Hello World` message. However, since the Jupyter server software is only expecting one user, there is only one copy of the x variable, so it displays its value.

JupyterHub

Once Jupyter Notebooks were shared, it became obvious that the multiuser problem had to be solved. A new version of the Jupyter software was developed called **JupyterHub**. JupyterHub was specifically designed to handle multiple users, giving each user their own set of variables to work with. Actually, the system will give each user a whole new instance of the Jupyter software to each user—a **brute-force** approach, but it works.

When JupyterHub starts, it begins a hub or controlling agent. The hub will start an instance of a listener or proxy for Jupyter requests. When the proxy gets requests for Jupyter, it turns them over to the hub. If the hub decides this is a new user, it will generate a new instance of the Jupyter server and attach all further interactions between that user and Jupyter to their own version of the server.

Installation

JupyterHub requires Python 3.3 or better, and we will use the Python tool `pip3` to install JupyterHub. You can check the version of Python you are running by just entering `Python` on a command line and the prologue will echo out the current version. For example, see the following:

```
Python
Python 3.6.0a4 (v3.6.0a4:017cf260936b, Aug 15 2016, 13:38:16)
[GCC 4.2.1 (Apple Inc. build 5666) (dot 3)] on darwin
Type "help", "copyright", "credits" or "license" for more information.
```

If you need to upgrade to a new version, refer to the instructions on `http://python.org` since they are specific to the operating system and version of Python.

JupyterHub is installed much like other software using the following commands:

```
npm install -g configurable-http-proxy
pip3 install jupyterhub
```

First, installing the proxy. `-g` on the proxy install means to make that software available to all users:

```
npm install -g configurable-http-proxy
/usr/local/bin/configurable-http-proxy ->
/usr/local/lib/node_modules/configurable-http-proxy/bin/configurable-http-
proxy
/usr/local/lib
└─┬ configurable-http-proxy@1.3.0
  ├─┬ commander@2.9.0
```

```
    |      └─── graceful-readlink@1.0.1
    ├──┬ http-proxy@1.13.3
    |  ├─── eventemitter3@1.2.0
    |  └─── requires-port@1.0.0
    ├──┬ lynx@0.2.0
    |  ├─── mersenne@0.0.3
    |  └─── statsd-parser@0.0.4
    ├─── strftime@0.9.2
    └──┬ winston@2.2.0
       ├─── async@1.0.0
       ├─── colors@1.0.3
       ├─── cycle@1.0.3
       ├─── eyes@0.1.8
       ├─── isstream@0.1.2
       ├─── pkginfo@0.3.1
       └─── stack-trace@0.0.9
```

Then, we install JupyterHub:

```
pip3.6 install jupyterhub
Collecting jupyterhub
  Downloading jupyterhub-0.6.1-py3-none-any.whl (1.3MB)
    100% |████████████████████████████████| 1.4MB
789kB/s
Collecting requests (from jupyterhub)
  Downloading requests-2.11.1-py2.py3-none-any.whl (514kB)
    100% |████████████████████████████████| 522kB
1.5MB/s
Collecting traitlets>=4.1 (from jupyterhub)
  Downloading traitlets-4.2.2-py2.py3-none-any.whl (68kB)
    100% |████████████████████████████████| 71kB
4.3MB/s
Collecting sqlalchemy>=1.0 (from jupyterhub)
  Downloading SQLAlchemy-1.0.14.tar.gz (4.8MB)
    100% |████████████████████████████████| 4.8MB
267kB/s
Collecting jinja2 (from jupyterhub)
  Downloading Jinja2-2.8-py2.py3-none-any.whl (263kB)
    100% |████████████████████████████████| 266kB
838kB/s
...
```

Operation

We can now start JupyterHub directly from the command line:

```
jupyterhub
```

This results in the following display that will appear in the command console window:

```
[I 2016-08-28 14:30:57.895 JupyterHub app:643] Writing cookie_secret to
/Users/dtoomey/jupyterhub_cookie_secret
[W 2016-08-28 14:30:57.953 JupyterHub app:304]
    Generating CONFIGPROXY_AUTH_TOKEN. Restarting the Hub will require
restarting the proxy.
    Set CONFIGPROXY_AUTH_TOKEN env or JupyterHub.proxy_auth_token config to
avoid this message.
[W 2016-08-28 14:30:57.962 JupyterHub app:757] No admin users, admin
interface will be unavailable.
[W 2016-08-28 14:30:57.962 JupyterHub app:758] Add any administrative users
to `c.Authenticator.admin_users` in config.
[I 2016-08-28 14:30:57.962 JupyterHub app:785] Not using whitelist. Any
authenticated user will be allowed.
[I 2016-08-28 14:30:57.992 JupyterHub app:1231] Hub API listening on
http://127.0.0.1:8081/hub/
[E 2016-08-28 14:30:57.998 JupyterHub app:963] Refusing to run JuptyerHub
without SSL. If you are terminating SSL in another layer, pass --no-ssl to
tell JupyterHub to allow the proxy to listen on HTTP.
```

Notice that the script completed, and a window did not open for you in your default browser as it would have in the standard Jupyter installation.

More important is the last line of output (which is also printed on screen in red), `Refusing to run JupyterHub without SSL`. JupyterHub is specifically built to account for multiple users logging in and using a single Notebook, so it is complaining that it is expected to have SSL running (to secure user interactions).

The last half of the last line gives us a clue as to what to do—we need to tell JupyterHub that we are not using a certificate/SSL. We can do that with the `--no-ssl` argument, as in the following:

```
Jupyterhub --no-ssl
```

This results in the expected outcome in the console, and leaves the server still running:

```
[I 2016-08-28 14:43:15.423 JupyterHub app:622] Loading cookie_secret from
/Users/dtoomey/jupyterhub_cookie_secret
[W 2016-08-28 14:43:15.447 JupyterHub app:304]
    Generating CONFIGPROXY_AUTH_TOKEN. Restarting the Hub will require
restarting the proxy.
    Set CONFIGPROXY_AUTH_TOKEN env or JupyterHub.proxy_auth_token config to
avoid this message.
[W 2016-08-28 14:43:15.450 JupyterHub app:757] No admin users, admin
interface will be unavailable.
[W 2016-08-28 14:43:15.450 JupyterHub app:758] Add any administrative users
```

```
to `c.Authenticator.admin_users` in config.
[I 2016-08-28 14:43:15.451 JupyterHub app:785] Not using whitelist. Any
authenticated user will be allowed.
[I 2016-08-28 14:43:15.462 JupyterHub app:1231] Hub API listening on
http://127.0.0.1:8081/hub/
[W 2016-08-28 14:43:15.468 JupyterHub app:959] Running JupyterHub without
SSL. There better be SSL termination happening somewhere else...
[I 2016-08-28 14:43:15.468 JupyterHub app:968] Starting proxy @
http://*:8000/
14:43:15.867 - info: [ConfigProxy] Proxying http://*:8000 to
http://127.0.0.1:8081
14:43:15.871 - info: [ConfigProxy] Proxy API at
http://127.0.0.1:8001/api/routes
[I 2016-08-28 14:43:15.900 JupyterHub app:1254] JupyterHub is now running
at http://127.0.0.1:8000/
```

If we now go to that URL shown (`http://127.0.0.1:8000/`) on the last line of the output, we get to a login screen for JupyterHub:

So, we have avoided requiring SSL, but we still need to configure the users for the system.

The JupyterHub software uses a configuration file to determine how it should work. You can generate a configuration file using JupyterHub, providing default values using the following command:

```
jupyterhub --generate-config
Writing default config to: jupyterhub_config.py
```

The generated configuration file has close to 500 lines available. The start of the sample file is as follows:

```
# Configuration file for jupyterhub.
c = get_config()


#-------------------------------------------------------------------------
----
# JupyterHub configuration
#-------------------------------------------------------------------------
----


# An Application for starting a Multi-User Jupyter Notebook server.
# JupyterHub will inherit config from: Application

# Include any kwargs to pass to the database connection. See
# sqlalchemy.create_engine for details.
# c.JupyterHub.db_kwargs = {}

# The base URL of the entire application

# c.JupyterHub.base_url = '/'
...
```

As you can see, most of the configuration settings are prefixed with a sharp (#) denoting that they are commented out. The setting that is mentioned is the default value that will be applied. If you needed to change one of the settings, you would remove the prefix sharp symbol and change the right-hand side of the equal sign (=) to your new value. By the way, this is a good way to test out changes: make one change; save the file; try out your change; continue with additional changes. As you progress, if one change does not work as expected, you need to just replace the prefix sharp symbol and you are back to a working position.

We will look at several of the configuration options available. It is interesting to note that many of the settings in this file are Python settings, not particular to JupyterHub. The list of items includes those shown here:

Area	Description
JupyterHub	Settings for JupyterHub itself
LoggingConfigurable	Logging information layout
SingletonConfigurable	A configurable that only allows one instance
Application	Date format and logging level
Security	SSL certificate
Spawner	How the hub starts new instances of Jupyter for new users
LocalProcessSpawner	Uses popen to start local processes as users
Authenticator	The primary API is one method, authenticate
PAMAuthenticator	Interaction with Linux to login
LocalAuthenticator	Checks for local users, and can attempt to create them if they exist

Continuing with operations

I made no changes to the configuration file to get my installation up and running. By default, the configuration uses the PEM system, which will hook into the operating system you are running on to pass in credentials (as if they were logging into the machine) for validation.

If you are seeing the `JupyterHub single-user server requires notebook >= 4.0` message in the console log when trying to log in to your JupyterHub installation, you need to update the base Jupyter using the command:

```
pip3 install jupyter
```

This will update your base Jupyter to the latest version, currently 4.1.

If you do not have `pip3` installed, you need to upgrade to Python 3 or better. See the documentation at `http://python.org` regarding the steps to be taken for your system.

Now, you can start JupyterHub using this following command line:

```
jupyterhub --no-ssl
```

Log in to the login screen (shown previously) using the same credentials you use to log in to the machine (remember JupyterHub is using PEM, which calls into your operating system to validate credentials). You will end up in something that looks very much like your standard Jupyter home page:

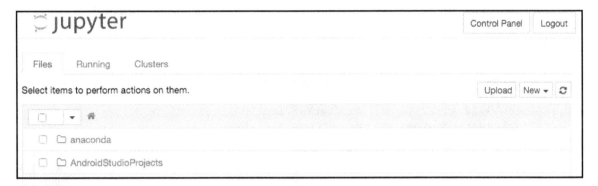

It looks very similar, except there are now two additional buttons in the top right of the screen:

- **Control Panel**
- **Logout**

Clicking on the **Logout** button logs you out of JupyterHub and redirects you to the login screen.

Clicking on the **Control Panel** button brings you to a new screen with two options, shown as follows:

- **Stop My Server**
- **My Server**

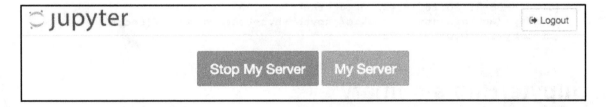

Clicking on the **Stop My Server** button stops your Jupyter installation and brings you to a page with one button: **My Server** (as shown in the following section). You may also have noticed the changes that have occurred in the console log of your command line:

```
[I 2016-08-28 20:22:16.578 JupyterHub log:100] 200 GET
/hub/api/authorizations/cookie/jupyter-hub-token-dtoomey/[secret]
(dtoomey@127.0.0.1) 13.31ms
[I 2016-08-28 20:23:01.181 JupyterHub orm:178] Removing user dtoomey from
proxy
[I 2016-08-28 20:23:01.186 dtoomey notebookapp:1083] Shutting down kernels
[I 2016-08-28 20:23:01.417 JupyterHub base:367] User dtoomey server took
0.236 seconds to stop
[I 2016-08-28 20:23:01.422 JupyterHub log:100] 204 DELETE
/hub/api/users/dtoomey/server (dtoomey@127.0.0.1) 243.06ms
```

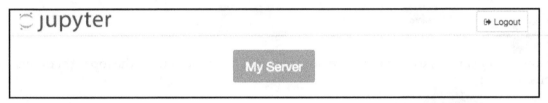

Clicking on the **My Server** button brings you back to the Jupyter home page. If you had hit the **Stop My Server** button earlier, then the underlying Jupyter software would be restarted, as you may notice in the console output (which I have shown as follows):

```
I 2016-08-28 20:26:16.356 JupyterHub base:306] User dtoomey server took
1.007 seconds to start
[I 2016-08-28 20:26:16.356 JupyterHub orm:159] Adding user dtoomey to proxy
/user/dtoomey => http://127.0.0.1:50972
[I 2016-08-28 20:26:16.372 dtoomey log:47] 302 GET /user/dtoomey
(127.0.0.1) 0.73ms
[I 2016-08-28 20:26:16.376 JupyterHub log:100] 302 GET /hub/user/dtoomey
(dtoomey@127.0.0.1) 1019.24ms
[I 2016-08-28 20:26:16.413 JupyterHub log:100] 200 GET
/hub/api/authorizations/cookie/jupyter-hub-token-dtoomey/[secret]
(dtoomey@127.0.0.1) 10.75ms
```

JupyterHub summary

So, in summary, with JupyterHub we have an installation of Jupyter that will maintain a separate instance of the Jupyter software for each user and thereby avoid any collision on variable values. The software knows whether to instantiate a new instance of Jupyter since the user logs in to the application and the system maintains a user list.

Docker

Docker is another mechanism that can be used to allow multiple users of the same Notebook without collision. Docker is a system that allows you to construct sets of applications into an image that can be run in a container. Docker runs in most environments. Docker allows for many instances of an image to be run in the same machine, but to maintain a separate address space. So, each user of a Docker image has their own instance of the software and their own set of data/variables.

Each image is a complete stack of software necessary to run, for example, a web server, web application(s), APIs, and more.

It is not a large leap to think of an image of your Notebook. The image contains Jupyter server code and your Notebook. The result is a completely intact unit that does not share any space with another's instance.

Installation

Installing Docker involves downloading the latest file (the `docker.dmg` file for a macOS and the `.exe` file install for Windows) and then copying the Docker applications into your `Applications` folder. **Docker QuickStart Terminal** is the go-to application of use to most developers. Docker QuickStart will start Docker on your local machine, allocate an IP address / port number for addressing the Docker application(s), and bring you into the Docker Terminal. Once QuickStart has completed, if you have installed your image, you could access your application (in this case, your Jupyter Notebook).

From the Docker Terminal, you can load images, remove images, check status, and more.

Starting Docker

If you run `Docker QuickStart`, you will be brought to the Docker Terminal window with a display like the following:

```
bash --login '/Applications/Docker/Docker Quickstart
Terminal.app/Contents/Resources/Scripts/start.sh'
Last login: Tue Aug 30 08:25:11 on ttys000
bos-mpdc7:Applications dtoomey$ bash --login '/Applications/Docker/Docker
Quickstart Terminal.app/Contents/Resources/Scripts/start.sh'

Starting "default"...
(default) Check network to re-create if needed...
```

```
(default) Waiting for an IP...
Machine "default" was started.
Waiting for SSH to be available...
Detecting the provisioner...
Started machines may have new IP addresses. You may need to re-run the
`docker-machine env` command.
Regenerate TLS machine certs?  Warning: this is irreversible. (y/n):
Regenerating TLS certificates
Waiting for SSH to be available...
Detecting the provisioner...
Copying certs to the local machine directory...
Copying certs to the remote machine...
Setting Docker configuration on the remote daemon...
```

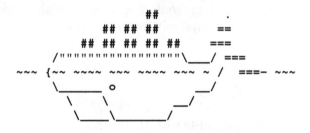

```
docker is configured to use the default machine with IP 192.168.99.100
For help getting started, check out the docs at https://docs.docker.com
```

(The odd graphic near the end of the display is a character representation of a whale—the logo for Docker.)

You can see the following from the output:

- The Docker machine was started—the Docker machine controls the images that are running in your space
- If you are using certificates, the certificates are copied into your Docker space
- Lastly, it tells you the IP address to use when accessing your Docker instances—it should be the IP address of the machine you are using

Building your Jupyter image for Docker

Docker knows about images that contain the entire software stack necessary to run an application. We need to build an image with a Notebook and place this in Docker.

We need to download all of the Jupyter-Docker coding necessary. In the Docker Terminal window, we run the following command:

```
$ docker pull jupyter/all-spark-notebook
Using default tag: latest
latest: Pulling from jupyter/all-spark-notebook
8b87079b7a06: Pulling fs layer
872e508604af: Pulling fs layer
8e8d83eda71c: Pull complete
...
```

This is a large download that will take some time. It is downloading and installing all of the possibly necessary components needed to run Jupyter in an image. Remember that each image is completely self-contained so each image has ALL of the software needed to run Jupyter.

Once the download is complete, we can start an image for our Notebook using a command such as the following:

```
docker run -d -p 8888:8888 -v /disk-directory:/virtual-notebook
jupyter/all-spark-notebook
The parts of this command are:
```

- `docker run`: The command to Docker to start executing an image.
- `-d`: Runs the image as a server (daemon) that will continue running until manually stopped by the user.
- `-p 8888:8888`: Exposes the internal port `8888` to external users with the same port address. Notebooks use port `8888` by default already, so we are saying just expose the same port.
- `-v /disk-directory:/virtual-notebook`: Takes the Notebook from the `disk-directory` and exposes it as the `virtual-notebook` name.
- The last argument is to use `all-spark-notebook` as the basis for this image. In my case, I used the following command:

  ```
  $ docker run -d -p 8888:8888 -v /Users/dtoomey:/dan-notebook
  jupyter/all-spark-notebook
  b59eaf0cae67506e4f475a9861f61c01c5af3556489992104c4ce39343e8eb02
  ```

The big hex number displayed is the image identifier. We can make sure the image is running using the `docker ps -l` command that lists out the images in our Docker:

```
$ docker ps -l
CONTAINER ID          IMAGE                          COMMAND
CREATED               STATUS               PORTS
NAMES
b59eaf0cae67          jupyter/all-spark-notebook     "tini -- start-
notebo"    8 seconds ago         Up 7 seconds
0.0.0.0:8888->8888/tcp    modest_bardeen
```

The parts of the display are as follows:

- The first name `b59...` is the assigned ID of the container. Each image in Docker is assigned to a container.
- The image is `jupyter/all-spark-notebook`, and it contains all of the software needed to run your Notebook.
- The command is telling Docker to start the image.
- The port access point is as we expected: `8888`.
- Lastly, Docker assigns random names to every running image `modest_bardeen` (not sure why they do this).

If we try to access Docker Jupyter at this point, we will be asked to set up security for the system, as in this display:

jupyter

Password or token: [   ] [Log in]

Token authentication is enabled

If no password has been configured, you need to open the notebook server with its login token in the URL, or paste it above. This requirement will be lifted if you enable a password.

The command:

```
jupyter notebook list
```

will show you the URLs of running servers with their tokens, which you can copy and paste into your browser. For example:

```
Currently running servers:
http://localhost:8888/?token=c8de56fa... :: /Users/yo
u/notebooks
```

or you can paste just the token value into the password field on this page.

See the documentation on how to enable a password in place of token authentication, if you would like to avoid dealing with random tokens.

Cookies are required for authenticated access to notebooks.

Setup a Password

You can also setup a password by entering your token and a new password on the fields below:

Token

[   ]

New Password

[   ]

[Log in and set new password]

Once we have set up security, we should be able to access the Notebook from a browser at `http:// 127.0.0.1:8888`. We saw the preceding IP address when Docker started (`0.0.0.0`) and we are using port `8888` as we specified:

You can see the URL in the top-left corner. Beneath that, we have a standard empty Notebook. The Docker image used has all of the latest versions, so you do not have to do anything special to get updated software or components for your Notebook. You can see the language options available by pulling down the **New** menu:

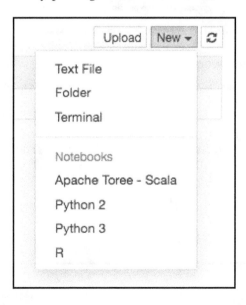

Docker summary

We have installed Docker, and we have created an image with our Notebook. We have also placed the Docker image into Docker, and we have accessed our Docker Notebook image.

Summary

In this chapter, we learned how to expose a Notebook so that multiple users can use a Notebook at the same time. We saw an example of an error occurring. We installed a Jupyter server that addresses the problem, and we used Docker to alleviate the issue as well.

In the next chapter, we will look at some upcoming feature enhancements for Jupyter.

12
What's Next?

In this chapter, we will do the following:

- Highlight some of the proposed upcoming features of Jupyter that you may look forward to. (Most of these examples are screenshots and descriptions of plans, without any coding available. The ideas presented are in the early stages and are not generally available)
- Give an example of an issue that occurs when multiple users access the same Notebook in a standard Jupyter installation.

JupyterHub

We saw JupyterHub in an earlier chapter.

JupyterHub is the multiuser instance of Jupyter. All directions where we need to have multiple users interact with a Jupyter implementation will be shown through Jupyter.

The first step was multiple users. This required Jupyter to allocate separate space for each user.

The next step is shared services that are accessed by the one Jupyter implementation, but many users. Services need to have authentication (is this really user xyzzy?) and Jupyter is moving toward using OAUTH, the standard for web service authentication.

JupyterLab

We have looked at JupyterLab in an earlier chapter.

JupyterLab is devoted to producing the next user interface design for Jupyter.

Scale

The current implementations of Jupyter have a scaling issue since, as the number of users increases, performance degrades significantly.

With the widespread adoption of Jupyter, the problem is becoming more apparent with additional use/users.

The Jupyter team is devoting a large part of their near-term efforts to enhancing the scalability of Jupyter to handle large numbers of users seamlessly. Some of the solutions being worked on include the following:

- A proxy API having an API between browsers and server services scales out the application
- MOAR servers, allowing multiple servers per user
- Further integration of OAUTH throughout the server/services
- Stress-testing work by the Jupyter team prior to market release

Custom frontends

With the extraction of the handling of user interface components, users can now have completely customized frontends for their implementation. All of the entry points are documented to perform the steps required.

Interactive computing standards

Jupyter is, by definition, an interactive computing platform. The Jupyter team is working with industry members to produce standards (that Jupyter will meet) in the areas of:

- Notebook file formats
- Interactive computing protocols
- Kernel interaction

For this chapter, we will use a simple Notebook that asks the user for some information, as demonstrated in the following screenshot:

Summary

In this chapter, we learned how to expose a Notebook so that multiple users can use a Notebook at the same time. We saw an example of the error occurring. We installed a Jupyter server that addressed the problem, and we used Docker to alleviate the issue as well.

Other Books You May Enjoy

If you enjoyed this book, you may be interested in these other books by Packt:

Hands-On Data Science with Anaconda
Dr. Yuxing Yan, James Yan

ISBN: 9781788831192

- Perform cleaning, sorting, classification, clustering, regression, and dataset modeling using Anaconda
- Use the package manager conda and discover, install, and use functionally efficient and scalable packages
- Get comfortable with heterogeneous data exploration using multiple languages within a project
- Perform distributed computing and use Anaconda Accelerate to optimize computational powers
- Discover and share packages, notebooks, and environments, and use shared project drives on Anaconda Cloud
- Tackle advanced data prediction problems

Jupyter Cookbook
Dan Toomey

ISBN: 9781788839440

- Install Jupyter and configure engines for Python, R, Scala and more
- Access and retrieve data on Jupyter Notebooks
- Create interactive visualizations and dashboards for different scenarios
- Convert and share your dynamic codes using HTML, JavaScript, Docker, and more
- Create custom user data interactions using various Jupyter widgets
- Manage user authentication and file permissions
- Interact with Big Data to perform numerical computing and statistical modeling
- Get familiar with Jupyter's next-gen user interface - JupyterLab

Leave a review - let other readers know what you think

Please share your thoughts on this book with others by leaving a review on the site that you bought it from. If you purchased the book from Amazon, please leave us an honest review on this book's Amazon page. This is vital so that other potential readers can see and use your unbiased opinion to make purchasing decisions, we can understand what our customers think about our products, and our authors can see your feedback on the title that they have worked with Packt to create. It will only take a few minutes of your time, but is valuable to other potential customers, our authors, and Packt. Thank you!

Index